T0258156

Analog Circuits Handbook

Analog Circuits Handbook

Edited by **Gus Winters**

New York

Published by NY Research Press,
23 West, 55th Street, Suite 816,
New York, NY 10019, USA
www.nyresearchpress.com

Analog Circuits Handbook
Edited by Gus Winters

© 2015 NY Research Press

International Standard Book Number: 978-1-63238-051-7 (Hardback)

Printed in the United States of America.

Contents

Preface   VII

Section 1   Circuit Design   1

Chapter 1   **Radio Frequency IC Design with Nanoscale DG-MOSFETs**   3
Soumyasanta Laha and Savas Kaya

Chapter 2   **A Successive Approximation ADC using PWM Technique for Bio-Medical Applications**   32
Tales Cleber Pimenta, Gustavo Della Colletta, Odilon Dutra, Paulo C. Crepaldi, Leonardo B. Zocal and Luis Henrique de C. Ferreira

Section 2   Analog CAD   48

Chapter 3   **Interval Methods for Analog Circuits**   50
Zygmunt Garczarczyk

Chapter 4   **Memetic Method for Passive Filters Design**   71
Tomasz Golonek and Jantos Piotr

Chapter 5   **Fault Diagnosis in Analog Circuits via Symbolic Analysis Techniques**   89
Fawzi M Al-Naima and Bessam Z Al-Jewad

Permissions

List of Contributors

Preface

In my initial years as a student, I used to run to the library at every possible instance to grab a book and learn something new. Books were my primary source of knowledge and I would not have come such a long way without all that I learnt from them. Thus, when I was approached to edit this book; I became understandably nostalgic. It was an absolute honor to be considered worthy of guiding the current generation as well as those to come. I put all my knowledge and hard work into making this book most beneficial for its readers.

This book discusses various aspects of analog circuits and its applications which will be beneficial for readers interested in this subject. Analog design explores innovative circuit topologies, architectures and CAD skills so as to conquer the design troubles emerging from new functions and new production technologies. In this book, a new structural design for a SAR ADC is suggested, so as to get rid of the procedure mismatches and reduce the errors. An analog circuit is one that uses nonstop time voltages and currents. This book is a collection of theories and researches based on analog circuits. It even explains radio frequency IC design, memetic method for passive filters design, among others.

I wish to thank my publisher for supporting me at every step. I would also like to thank all the authors who have contributed their researches in this book. I hope this book will be a valuable contribution to the progress of the field.

Editor

Circuit Design

Radio Frequency IC Design with Nanoscale DG-MOSFETs

Soumyasanta Laha and Savas Kaya

Additional information is available at the end of the chapter

1. Introduction

Today's nanochips contain billions of transistors on a single die that integrates whole electronic systems as opposed to sub-system parts. Together with ever higher frequency performances resulting from transistor scaling and material improvements, it thus become possible to include on the same silicon chip analog functionalities and communication circuitry that was once reserved to only an elite class of compound III-V semiconductors. It appears that the last stretch of Moore's scaling down to 5 nm range, only limited by fabrication at atomic dimensions and fundamental physics of conduction and insulation, these systems will only become more capable and faster, due to novel types of transistor geometries and functionalities as well as better integration of passive elements, antennas and novel isolation approaches. Accordingly, this chapter is an example to how RF-CMOS integration may benefit from use of a novel multi-gate transistors called FinFETs or double-gate MOSFETs (DG-MOSFETs). More specifically, we hope to illustrate how radio frequency wireless communication circuits can be improved by the use of these novel transistor architectures.

1.1. CMOS downscaling to DG-MOSFETs

As device scaling aggressively continues down to sub-32nm scale, MOSFETs built on Silicon on Insulator (SOI) substrates with ultra-thin channels and precisely engineered source/drain contacts are required to replace conventional bulk devices [1]. Such SOI MOSFETs are built on top of an insulation (SiO_2) layer, reducing the coupling capacitance between the channel and the substrate as compared to the bulk CMOS. The other advantages of an SOI MOSFET include higher current drive and higher speed, since doping-free channels lead to higher carrier mobility. Additionally, the thin body minimizes the current leakage from the source to drain as well as to the substrate, which makes the SOI MOSFET a highly desirable device applicable for high-speed and low-power applications. However, even these redeeming features are not expected to provide extended lifetime for the conventional MOSFET scaling below 22nm and more dramatic changes to device geometry, gate electrostatics and channel material are required. Such extensive changes are best introduced gradually, however, especially when it comes to new materials. It is the focus on 3D transistor geometry and electrostatic design, rather than novel materials, that make the multi-gate (i.e double, triple, surround) MOSFETs as one of the most suitable candidates for the next phase of evolution in Si MOSFET technology   [2]- [5].

Being the simpler and relatively easier to fabricate among the multigate MOSFET structures (MIGFET, Π-MOSFET and so on) the double gate MOSFET (DG-MOSFET) (Fig. 1) is chosen here to explore these new circuit possibilities. The DG-MOSFET architectures can efficiently control the channel from two sides of instead of one as in planar bulk MOSFETs. The advantages of DG-MOSFETs are as follows [6]:

- Reduced Short Channel Effects (SCE) due to the presence of two gates and ultra-thin body.
- Reduced subthreshold leakage current due to reduced SCE.
- Reduced gate leakage current due to the use of thicker oxide. Lower SCE in DG devices and the higher driver current (due to two gates) allows the use of thicker oxide in DG devices compared to bulk-CMOS structures.

Due to the reasons stated above, the last decade has witnessed a frenzy of design activity to evaluate, compare and optimize various multi-gate geometries, mostly from the digital CMOS viewpoint [7], [8]. While this effort is still ongoing, the purpose of the present chapter is to underline and exemplify the massive increase in the headroom for CMOS nano-circuit engineering of RF communication systems, when the conventional MOSFET architecture is augmented with one extra gate.

The great potential of DG-MOSFETs for new directions in tunable analog and reconfigurable digital circuit engineering has been explored before in [9]. The innate capability of this device has also been explored by others, such as the Purdue group led by K. Roy [6], [7] has demonstrated the impact of DG-MOSFETs (specifically in FinFET device architecture) for power reduction in digital systems and for new SRAM designs. Kursun (Wisconsin & Hong Kong) has illustrated similar power/area gains in sequential and domino-logic circuits [10]. A couple of French groups have recently provided a very comprehensive review of their DG-MOSFET device and circuit works in a single book [8]. Their works contain both simulation and practical implementation examples, similar to the work carried out by the AIST XMOS and XDXMOS initiative in Japan   [11]-[13] as well as a unique DG-MOSFET implementation named FlexFET by the ASI Inc [14], [15]. Recently, Intel has announced the most dramatic change to the architecture of the transistor since the device was invented. They will henceforth build transistors in three dimensions, which they called the 3D-MOSFET [4], a device that corresponds to FinFET/DG-MOSFET.

1.2. RF/Analog IC design

In addition to features essential for digital CMOS scaling such as the higher I_{ON}/I_{OFF} ratio and better short channel performance, DG-MOSFETs possess architectural features also

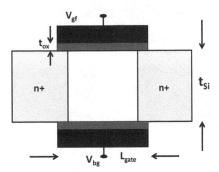

Figure 1. Generic DG MOSFET structure.

helpful for the design of massively integrated radio frequency analog and adaptive systems with minimal overhead to the fabrication sequence. Given the fact that they are designed for sub-22nm technology nodes, the DG MOSFETs can effectively handle GHz modulation, making them relevant for the RF/Analog/Mixed-Signal system-on-chip applications and giga-scale integration [16], [17].

The two most important metrics for RF CMOS/DG-CMOS circuits are the transit frequency f_T and the maximum oscillation frequency f_{max}. The former is defined as the frequency at which the current gain of the active device is unity, while the latter is the frequency for which the power gain is unity. Both these quantities relate the achievable transconductance to "parasitics" as gate-source and gate-drain capacitances (C_{gs} and C_{gd}). In case of f_{max} the gate resistance R_G is also considered as it deals with power dissipation. The f_T increases with decreasing gate lengths and for a DG-MOSFET at 45 nm it is obtained around 400 GHz [18].

Also, they have reduced cross-talk and better isolation provided naturally by the SOI substrate, multi-finger gates, low parasitics and scalability. However, the DG-MOSFET's potential for facilitating mixed-signal and adaptive system design is highest when the two gates are driven with independent signals [19]. It is the independently-driven mode of operation that furnishes DG MOSFET with a unique capability to alter the front gate threshold via the back gate bias. This in turn leads to:

• Increased operational capability out of a given set of devices and circuits.

• Reduction of parasitics and layout area in tunable or reconfigurable circuits.

• Higher speed operation and/or lower power consumption with respect to the equivalent. conventional circuits.

2. DG MOSFET modeling and simulation

2.1. ASU PTM for FinFETs

The widely available compact models for SOI-based single-gate MOSFETs can not be used for the DG-MOSFETs, for which new surface-potential based models are developed   [20]-[23].

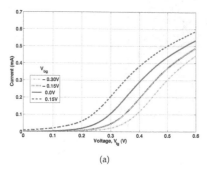 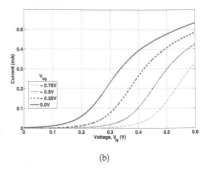

(a)   (b)

Figure 2. The I_D - V_{fg} characteristics of an n-type DG-MOSFETs at different back-gate bias conditions as observed in a) ASU PTM 32 nm DG FinFET b) ASU PTM 45 nm DG-FinFET technology with Synopsys HSPICE RF simulation.

Instead either physically-rigorous demanding TCAD simulations or approximate SPICE models utilizing two back-to-back MOSFETs mathematically coupled for improved accuracy may be used. In this chapter, most of the circuits investigated use this latter approach. We have used the ASU Predictive Technology Model for 45 nm & 32 nm DG FinFETs [24] for our simulations for most of the circuits. The circuit simulator used for the design and analysis is the industry standard Synopsys HSPICE RF. The reliability of these two ASU technology models are evident from the typical transfer characteristics of an n-type DG-MOSFET with independent back-gate biasing as shown in Figs. 2a & b. It is obvious that the front gate threshold can be tuned via the applied back-gate voltage, which is sufficient for us to confirm the tunable functionality and carry out a comparative study. This 'dynamic' threshold control is crucial to appreciate the tunable properties of the oscillator and amplifier circuits.

2.2. UFDG SPICE

The UFDG model is a process/physics and charge based compact model for generic DG MOSFETs [25]. The key parameters are related directly to the device physics . This model is a compact Poisson-Schrodinger solver for DG MOSFETs that physically accounts for the charge coupling between the front and the back gates. The UFDG allows operation in the independent gate mode and is applicable to FD SOI MOSFETs. The quantum mechanical modeling of the carrier confinement, dependent on the Ultra Thin body (UTB) thickness (t_{Si}) as well as transverse electric field is incorporated via Newton Raphson iterations that link it to the classical formalism.

The dependence of carrier mobility on Si-film thickness, subject to the QM confinement and on transverse electric field is also accounted for in the model. The carrier velocity overshoot and dependence on carrier temperature is characterized in the UFDG transport modeling to account for the ballistic and quasiballistic transport in scaled DG MOSFETS [26]. The channel current is limited by the thermal injection velocity at the source, which is modeled based on the QM simulation. The UFDG model also accounts for the parasitic (coupled) BJT (current and charge) which can be driven by transient body charging current (due to capacitive coupling) and/or thermal generation, GIDL [27] and impact ionization currents, the latter of which is characterized by a non-local carrier temperature-dependent model for the ionization rate integrated across the channel and the drain.

The charge modeling which is patterned after that is physically linked to the channel-current modeling. All terminal charges and their derivatives are continuous for all bias conditions, as are all currents and their derivatives. Temperature dependence for the intrinsic device characteristics and associated model parameters are also implemented without the need for any additional parameters. This temperature dependence modeling is the basis for the self-heating option, which iteratively solves for local device temperature in DC and transient simulations in accord with a user defined thermal impedance.

The Relaxation Oscillator and the RF-Mixer analysis are carried with this simulator.

3. Transmitter design

The transmitter (Fig. 3) consists of an oscillator, modulator, power amplifier and finally an antenna. A matching network (Z_0 in Fig. 3) which maximizes the power transfer and minimizes the reflection losses generally precedes the 50 Ω antenna. In this article, the components that have been investigated with DG-MOSFET technology include a Relaxation Oscillator, LC Oscillator, an OOK Modulator and two different topologies of Power Amplifier. It is to be noted that the oscillators are also part of the receiver design and has its use in RF Mixer and Phase Locked Loops (PLLs).

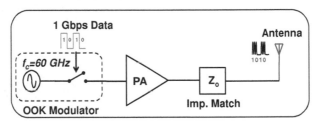

Figure 3. The transmitter block consisting of the oscillator, modulator and power amplifier and other passive devices/circuits.

3.1. Relaxation oscillator

Relaxation oscillator is an inductorless non-resonant oscillator that is either current controlled or voltage controlled. The second circuit in [28] implements a dual input S/R latch. As illustrated in Fig. 4a the NOR gates used to construct the latch consist of only four DG-MOSFET as opposed to eight required in conventional CMOS architecture. This serves to save circuit area and a decent amount of power dissipation. The two inverters are biased with equal copies of the input current, I_{IN}, from the current mirrors implemented with three pMOS. The back gate of the two inverters are tuned in voltage to vary the frequency.

The DG-MOSFET implementation also has two advantages, firstly it can be used also as a VCO by virtue of the back gate bias and secondly it operates more efficiently with a higher upper limit as a result of very high transconductance of DG-MOSFETs [29]. Although the accessible frequency range in the VCO mode is dwarfed in contrast to massive ICO response given in logarithmic scale, the operation as a VCO provides the circuit with an extra degree of freedom in tuning. Specifically, the voltage operated fine 'vernier' frequency tuning sets a frequency with precision after it has been 'coarsely' selected by the current operated crude logarithmic tuning.

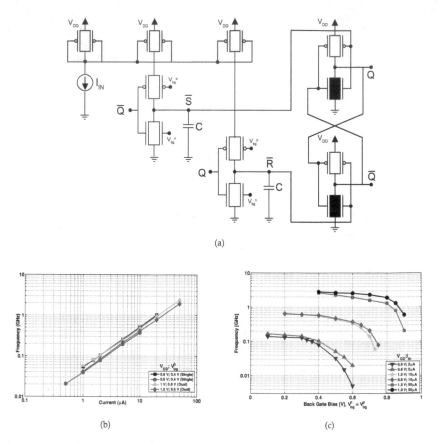

(a)

(b)   (c)

Figure 4. a)The current/voltage controlled relaxation oscillator in DG-MOSFET technology. b) The 'crude tuning' of the relaxation oscillator with varying current. c) The fine tuning in frequency with back gate bias when $V_{bg}^p = V_{bg}^n$ of the relaxation oscillator.

In Fig. 4b, we can verify the frequency has a log-log relationship with the current. The frequency ranges from 30 MHz to a few GHz for a change in current supply from 0.4 μA to 50 μA. This coarse tuning in frequency is supported via back gate fine tuning of the DG MOSFET inverters. For a constant current and voltage supply, the frequency can be tuned to vary in the order of MHz, as the inverter back gate voltage varies from 0.1 V to 1 V. It is observed, a higher V_{DD} results in a slower oscillation at a fixed input current, because the SR Latch takes longer time to reach a higher switching threshold ($\sim 1/2V_{DD}$) as V_{DD} is increased. The Fig. 4c demonstrates these facts with three different current sources and supply voltage. The phase noise of the oscillator is -104 dBc/Hz at 1 MHz offset. All these analysis are carried with 45 nm DG-MOSFET using UFDG SPICE.

3.2. LC oscillator and OOK modulator

LC oscillators consists of inductors and capacitors connected in parallel. Although inductors consume a lot of area when compared to the inductorless oscillator described above oscillators, it is a must in RF Design to use inductors because of two primary reasons [30]. They are as follows:

- The resonance of inductors with capacitors allow for higher operational frequency and lower phase noise.

- The inductor sustains a very small DC voltage drop which aids in low supply operation.

We have chosen the differential negative resistance voltage controlled oscillator (VCO) variant of the LC oscillator (Fig. 5a) for the investigation. The latch circuit in the differential mode serves as negative resistance to nullify the effects of a positive resistance arising out of the imperfect inductor. The Q factor determines the undesired resistance value (R) of the inductor (L) at the resonance frequency, ω. Modeling the resistive loss in the inductor, L by the parallel resistance (R) we can write [30]:

$$Q = \frac{R}{\omega L} \tag{1}$$

The LC tank achieves a frequency that is much higher and has a phase noise that is much lower than that of the relaxation oscillator. This is primarily because of the resonance of the circuit.

The OOK Modulation is a non-coherent modulation scheme that modulates the carrier only when the circuit is in the 'ON' state. It is the special case of Amplitude Shift Key (ASK) modulation where no carrier is present during the transmission of a 'zero'. The bit error rate for OOK modulation without the implementation of any error correcting scheme is given by [31]

$$BER = \frac{1}{2}exp(\frac{-E_b}{2N_0}) + \frac{1}{2}Q\sqrt{\frac{E_b}{N_0}} \tag{2}$$

Although, the associated bit error rate of OOK modulation is inferior to that of other coherent modulation schemes, simple OOK modulation scheme is implemented to avoid the complicated carrier recovery circuit and for their ability to modulate very high frequency signals in extremely long-life battery operated applications. The non-coherent OOK demodulation generally employs an envelope detector in the receiver which saves the power, area, cost and complexity since no local oscillator (LO) or carrier synchronization scheme is involved.

3.2.1. Design and simulation

The DG-MOSFET based VCO can be tuned from the back gate for controlling the rms voltage (V_{rms}). Fig. 5b illustrates this interesting tunable feature of the DG MOSFET VCO. Without any change in the supply, the V_{rms} can be controlled via back gate bias (V_{bg}), which can have

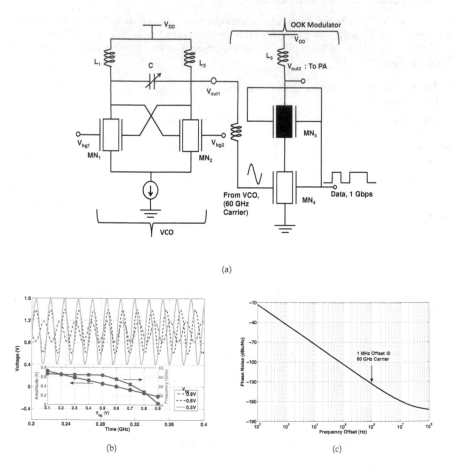

(a)

(b)   (c)

Figure 5. a) The OOK Modulator circuit with the VCO. The proposed OOK Modulator uses only two DG-MOSFET for modulation and switching. b) The variation of VCO output amplitude at different V_{bg}. Inset: Amplitude and frequency variation for different V_{bg}. c) The phase noise of the VCO at 60 GHz. The phase noise at 1 MHz offset is observed at -133 dBc/Hz in time variant Hajimiri-Lee model [32].

application in many adaptive low power wireless systems. The bias at the back gate can also be tuned to change the oscillation frequency after a certain threshold (0.5 V) (Fig. 5b inset). Although DG-MOSFET is not reputed for its noise performance, the phase noise of the 60 GHz VCO is found to be -133 dBc/Hz at 1 MHz offset (Fig. 5c) which is comparable to that of bulk CMOS [33]. As expected, the phase noise is dominated by the process dependent flicker noise of slope $\sim$ -30 dB/decade. The corner frequency f_{cor} is obtained around 10 MHz.

The proposed novel DG-MOSFET based OOK Modulator [34] consists of only two DG-MOSFETs making it ideal for use in ultra low power systems (Fig. 5a). The modulator can

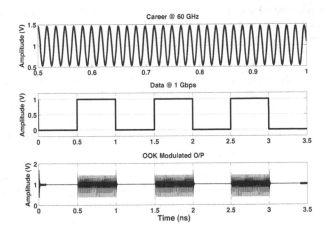

Figure 6. The OOK Modulated output for a carrier frequency of 60 GHz and data rate of 1 Gbps. The input data sequence resembles 50% duty cycle.

work up to a data rate of 5 Gbps without any discernible distortion for 60 GHz carrier. The DG-MOSFET MN_4 acts as the key OOK modulating device. The 60 GHz sinusoidal carrier from the VCO is fed into one of the gates of the transistor whereas the pulsed digital data is input to the other gate. The charge capacitive coupling of the two gates provided by the thin Si body determines the modulation, and therefore depends on the bias conditions of the two gates as well as device dimensions. The modulation occurs when the device operates in the saturation or in cut-off region, that is when there is either a '1' or '0' respectively emanating from the pulsed digital data. In other words, the modulation takes place at all instants of time. The symmetric DG-MOSFET MN_3 acts as the switch and is kept at a high threshold voltage (filled) for better electrostatics and keying and to maximize the I_{ON}/I_{OFF} ratio. The MN_3 is turned on at the 'HIGH' state of the pulsed data and remains off at the 'LOW' state, maintaining the principle of OOK Modulation scheme. The modulated output is obtained at the drain of MN_3. This is illustrated in Fig. 6. All these analysis are carried in 32 nm ASU PTM FinFET technology.

3.3. Power Amplifier

The Power Amplifier (PA) is the final stage of transmitter design before signal transmission through antenna. They are responsible for amplifying the power level of the transmitted signal several times so that the received signal is above the sensitivity of the receiver which is calculated from the link budget analysis. The PAs are divided into various classes such as A, B, AB, C D, E, F etc. Among these classes A, B, AB and C incorporate similar design methodologies differing only in the biasing point. Among these Class A amplifier is the most linear and is widely used in RF transmitter design although they have the least Power Added Efficiency (PAE). Several acclaimed literatures [35], [30] are available for interested readers on these concepts. This book chapter focusses on the design of tunable DG-MOSFET Class A PA.

The design of the wide band and high gain PA is a challenging task, especially in ultra-compact MOSFETs with low output impedance. Consequently, in [36], we simply adapted two recent single-gate implementations with competitive features in the GHz range, which allows a more fair performance comparison to be made between different devices. In the first PA topology [37], we modify the architecture slightly for the DG-MOSFET to explore its gain and bandwidth characteristics as well as its tunability. The second topology reported here is a three stage single-ended, common-source (CS) PA similar to the one reported by Yao et al. [38] for conventional CMOS. The basic difference over the published topologies in both cases is the length of the DG-MOSFET devices (45 nm) that is substantially smaller. There are a number of reasons for this gate length choice. Firstly, the proposed PAs are essentially designed for low-power highly compact Si mixed-signal radio applications where the range and area will be typically quite limited. Secondly, the DG-MOSFET architecture is inherently a narrow width device technology in which very large number of fingers needed to obtain large W/L ratios. Finally, we wish to implement a PA for ultra-compact wide-band RF CMOS applications such as vehicular anti-collision radar. Given that DG-MOSFET technology is aimed for sub-22 nm digital technologies, 45 nm is a good compromise for analog circuit implementation.

The next two sections will discuss in detail about these design modifications and provide their simulated response including gain tuning, peak gain, bandwidth and linearity. Interested readers can compare the performances of these power amplifiers with a few other conventional designs in [36].

3.3.1. Topology A - Design and simulation

The circuit topology of the first wide band (3-33 GHz) DG-MOSFET PA is shown in Fig. 7a, which consists of three DG-MOSFETs in a Darlington cascode arrangement. The common source transistor MN_1 operates in the symmetric mode while the two transistors MN_2 and MN_3 are configured for independent mode operation. The width of MN_1 is taken to be 1 μm while the width for transistors MN_2 and MN_3 are kept higher at 2.4 μm for better input return loss and optimized gain performance. MN_3 is biased at 2.6 V (V_{b1}). The back gate of the transistors MN_2 and MN_3 are biased for gain tuning. The resistors R_1 and R_2 complete a self biasing network for Class A operation. This modified DG-MOSFET darlington configuration is divided into two stages. The first stage is the series peaking stage and inter-stage matching, and the second stage is the output power stage.

The series peaking circuit consisting of R_3 and L_1 increases the output load pull impedance, and also provides the peaking impedance for feeding forward signals. The inductor L_3 along with the source degeneration circuit consisting of R_4 and L_2 yields in real part wide band inter-stage impedance matching for maximizing the power transfer between the stages. The common source transistor MN_2 and MN_3 are connected in cascode. The transistor MN_3 acts in common gate configuration and one of its gate is grounded with the aid of the peaking inductor L_4 and a bypass capacitor C_1 [30]. Along with achieving a near constant gain by maintaining the flatness, the bandwidth of the amplifier is also increased with the aid of this peaking inductor. A high pass L-network (L_5 & C_3) is used as the matching circuit.

Our simulation verifies the forward gain (S_{21}) to vary from 3 to 33 GHz, while maintaining a desired flatness (Fig. 7b). The gain changes by less than 20% in this frequency range, attesting to the extreme flatness. The peak gain is observed at 24.5 dB. The input and output return

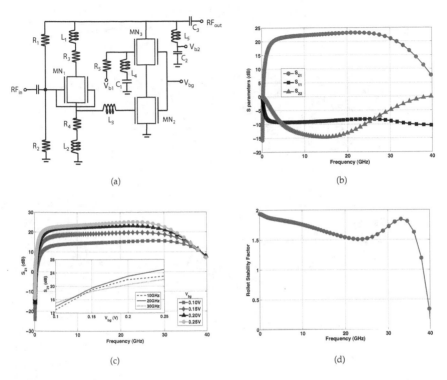

(a)   (b)

(c)   (d)

Figure 7. a) The DG MOSFET based power amplifier circuit in modified darlington cascode configuration. Transistors MN_1 operates in the symmetric mode while MN_2 and MN_3 operate in independent mode with the back gates used for dynamic tuning. b) The S parameters which provide the gain (S_{21}) and reflection losses (S_{11} & S_{22}) of the power amplifier. This is measured for $V_{bg} = 0.2$ V. c) The back gate dependence of the gain is clearly evident. The gain changes by ~ 10 dB in the tuning range of V_{bg}. Inset: Gain variation with V_{bg} at different frequencies. d) The rollet stability factor (K) is above unity in the operating range of 2 - 32 GHz verifying the amplifier to remain unconditionally stable in this range. K drops below unity beyond ~ 38 GHz.

losses (S_{11} & S_{22}) are also obtained from the simulation. Fig. 7c shows these S parameters at a V_{bg} of 0.2 V which is applied at the back gate of the transistors MN_2 and MN_3. The back gate voltage (V_{bg}) is varied from 0.1 V to 0.25 V for the operating frequency range during which the gain of the amplifier increases considerably. The range of gain tuning is observed to be limited to almost 10 dB. The inset of the figure shows the gain variation with V_{bg} at different frequencies. The unconditional stability of the amplifier is verified measuring the rollet stability factor, K which is given as

$$K = \frac{1 - |S_{11}|^2 - |S_{22}|^2 + |\triangle|^2}{2|S_{12}S_{21}|} \qquad (3)$$

$$\triangle = S_{11}S_{22} - S_{12}S_{21} \qquad (4)$$

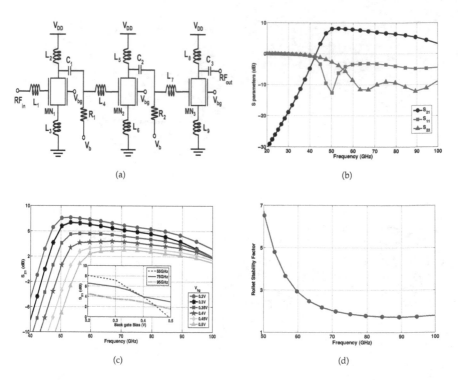

Figure 8. a) The three stage DG MOSFET based power amplifier circuit. All the three transistors operate in the independent mode. b) The S parameters which provide the gain (S_{21}) and reflection losses (S_{11} & S_{22}) of the power amplifier. This is also measured for V_{bg} = 0.2 V. c) The back gate dependence of the gain is clearly evident. The gain changes by $\sim$ 6 dB in the tuning range of V_{bg}. Inset: Gain variation with V_{bg} at different frequencies. d) The rollet stability factor (K) is well over unity in the operating range of 60 - 90 GHz verifying the amplifier to remain unconditionally stable in the range.

The value of K is observed to be above unity in the operating frequency range indicating the unconditional stability of the amplifier (Fig. 7d). The back gate tuning of the PA is verified from Fig. 5. The 1 dB compression point (P_{1dB}) and the 3rd order Input Intercept Point (IIP_3) are found to be 11.9 dBm and 27.5 dBm, respectively, indicating the suitability of the circuit. The 15.6 dB difference between P_{1dB} and IIP_3 can be attributed to the scaling down of DG MOSFET to 45 nm [35]. The power added efficiency (PAE) and the fractional bandwidth (FB) of the amplifier is $\sim$12% and 176% respectively.

3.3.2. Topology B - Design and simulation

In the second topology, the DG-MOSFET Class A amplifier is implemented in three stages (Fig. 8a). Although the earlier cascode topology has higher & flatter gain, and larger output impedance, the CS configuration is advantageous in terms of the lower supply voltage required, leading to higher efficiency. All the transistors in this topology operate in the

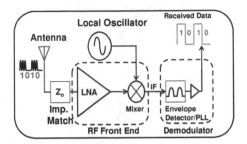

Figure 9. The receiver block consisting of the RF Front End (LNA & RF Mixer) and the Demodulator (Envelope Detector, for non-coherent detection or PLL, for coherent detection).

independent mode. The source degeneration inductors L_3, L_6 and L_9 along with the inter stage inductors L_4 and L_7 maximizes the power transfer and improves linearity [35]. The width of the three transistors are kept fixed at 1.2 μm. The source and the bias voltage (V_b) are both kept at 1 V.

Although the 3-dB bandwidth is $\geq$ 50 GHz, as evident from Fig. 8b, for all cases of back gate voltages (Fig. 8c) a more realistic operating range of this amplifier can be considered to be in the range of 60 - 90 GHz. Once again, the inset of the Fig. 8 shows the gain variation with V_{bg} at different frequencies. The peak gain achieved is $\geq$ 8 dB. The rollet stability factor remains more than unity for this operating range as shown by simulated data in Fig. 8d. The P_{1dB} and the IIP$_3$ are found to be 7.2 dBm and 19.8 dBm respectively. The PAE and the FB of this amplifier is $\sim$14% and 40% respectively.

4. Receiver design

The front end of the receiver consists of a Low Noise Amplifier (LNA) and RF Mixer. To demodulate a non-coherent signal an Envelope Detector is used while to demodulate a coherent signal a Phase Locked Loop is generally used (Fig. 9). In this chapter, we have designed an LNA, Envelope Detector and a Charge Pump Phase Frequency Detector (which is an essential component in PLL design) and analyzed an existing RF Mixer.

4.1. Low Noise Amplifier

The Low Noise Amplifier (LNA) is an essential component in the front-end of any communication/navigation receiver. The received signal at antenna is very weak and therefore it is necessary to amplify the signal for demodulation and processing. At the same time the noise figure of the amplifier has to be very low because the received signal will eventually be passed to non-linear devices such as RF Mixers which add noise. Therefore LNA design optimizes to minimize the noise level at the first stage of the receiver i.e. at the LNA itself. Other characteristics that require from an LNA include high gain, impedance matching linearity and stability.

The circuit topology of the tunable 45 nm DG-MOSFET LNA implemented here is shown in Fig. 10, which consists of three DG-MOSFETs in a 2 stage common source cascode topology. The common source transistor MN_1 operates in the symmetric mode while the two transistors MN_2 and MN_3 are configured for independent mode operation. The common

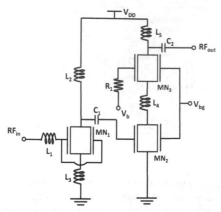

Figure 10. The DG MOSFET based LNA in common source cascode configuration. Transistors MN_1 operates in the symmetric mode while MN_2 and MN_3 operate in the independent mode with the back gates used for dynamic tuning.

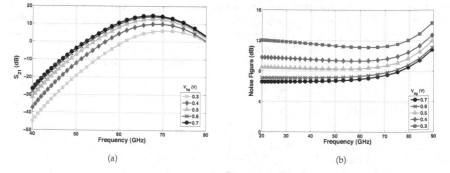

Figure 11. a) The gain (S_{21}) of the LNA varies with V_{bg}. This is measured for V_{bg} = 0.3 V to 0.7 V. b) The noise figure dependence on V_{bg} of the LNA is evident. The NF changes by 4.4 dB in the tuning range of V_{bg} at 65 GHz.

source transistor MN_2 and MN_3 are connected in cascode. The transistor MN_3 acts in common gate configuration. The width of MN_1 is taken to be 1 μm while the width for transistors MN_2 and MN_3 are kept higher at 2.4 μm for better input return loss and optimized gain performance. The supply, V_{DD} is kept constant at 1.2 V. MN_3 is biased at 2 V (V_b). The back gate of the transistors MN_2 and MN_3 are biased for gain tuning.

The series peaking circuit consists of an inductive load, L_2, that allows for low voltage operation and resonates with the inter stage capacitance, C_1, enabling a higher operating frequency [30]. The inductor L_1 is set to resonate with the gate source capacitance of MN_1. The source degeneration circuit consisting of L_3 yields (in real part) wide band impedance matching to maximize the inter-stage power transfer. The inductor L_4 tunes out the middle pole of the cascode, thus compensating for the lower f_T [39] of DG-MOSFET which is nearly 150 GHz at 45 nm.

The simulation shows the 3-dB bandwidth to be 15 GHz, ranging from 60 to 75 GHz. The forward gain (S_{21}) achieves a peak value of 15 dB at 65 GHz for V_{bg} = 0.7 V (Fig. 11a). Beyond this maximum operating voltage the gain gets saturated and is independent of V_{bg}. The peak gain reduces gradually as V_{bg} is reduced and drops to ~5 dB for V_{bg} = 0.3 V. The power dissipated (P_{dc}) by the LNA also varies with V_{bg}, reaching 18 mW at V_{bg} = 0.7 V. Similarly, the LNA noise figure (NF) also depends upon the back gate bias, dropping to a minimum at peak gain as expected. It ranges from 7 dB at V_{bg} = 0.7 V to 11.4 dB at V_{bg} = 0.3 V (Fig. 11b). Clearly, the back gate tuning provides a convenient tool to optimize specific device performance parameters, setting up unique trade-offs such as that between power and gain.

The proposed LNA is unconditionally stable in the operating frequency range, verified from the simulated rollet stability factor, i.e. K > 1. The circuit is also simulated for linearity performance using a two tone frequency analysis near 60 GHz and the observed 3rd order Input Intercept Point (IIP3) is −5.2 dBm.

Overall, the DG-MOSFET implementations have impressive characteristics that either match or exceed the bulk MOSFET and even SiGe counterparts [40]-[42]. It is fair to point out that much of this response can be attributed to small gate length in our designs. However, a short gate length has also consequences for linearity and lower output impedance, with which this architecture appears to cope well.

4.2. RF mixers

The RF mixer is a non-linear electrical circuit that creates two new frequencies from the two signals applied to it. The new frequencies (sum & difference) are called the intermediate frequencies (IF). The sum frequency has its application on the up conversion whereas the difference frequency is used in the down conversion of an input signal. The conversion gain (CG) determines the mixing performance of the circuit [35].

4.2.1. DG-MOSFET mixers and methodology

DG-MOSFET mixer occupies a special status among analog applications given the compact and high performance nature of an active mixer using only one transistor which saves both power and area compared to conventional CMOS. Accordingly, there are already several literatures available focusing solely on this simple but promising circuit. For instance, a recent work by S. Huang et. al. [43] analyzes the RF Mixer based on the derivative superposition method. An earlier work [44] considers the evaluation of power consumption and area overhead of the DG-MOSFET for RF-mixer applications. W. Zhang et. al. [45] explored the use of multiple independent-gate FinFETs (MIGFETs) application and compares the spectral response of the single-and multiple-transistor (balanced) versions. Although this research provides valuable physical insights regarding the operational principles and behavior of the DG-MOSFET mixer, unfortunately the temporal resolution or the length of the transient data used in their Fast Fourier Transform (FFT) analysis, and the range of device parameters explored, are insufficient for a thorough study of the mixing performance in a methodical manner.

In contrast, in [46] we focus on the structural and operational parameters of DG-MOSFET in a methodical and accurate manner to optimize the biasing for maximum conversion

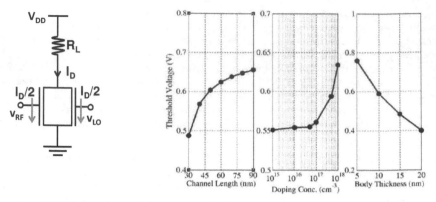

Figure 12. a) DG MOSFET RF Mixer circuit. b) The variation of V_T for the different device parameters L_{gate}, N_d and t_{Si}.

gain and power efficient design of the mixer circuit. Additionally, we also look into the correlation between conversion gain and the device parameters. In our methodology, we carefully considered the input RF and local oscillator (LO) signals' bias conditions while fairly comparing conversion gain recorded with different structural parameters, ensuring that gate over-drive (LO DC offset) has been kept the same.

The DG-MOSFET RF Mixer circuit (Fig. 12a) explored in [46] consists of a single double gate transistor. For a fair comparison of mixing performance obtained from varying important structural parameters, we first explore the dependence of threshold voltage (V_T) (Fig. 12b) on each of the device parameters, gate length (L_g), doping concentration (N_A) and body thickness (t_{Si}). The source voltage (V_{DD}) is kept at the typical value of 1 V and the circuit load, R_L, is kept at 6 kΩ for the analysis. The sinusoidal RF signal is considered at the frequency f_{RF} of 50 MHz while the sinusoidal local oscillator signal is chosen at a frequency f_{LO} of 10 MHz so to down convert the incoming frequency to 40 MHz. In this DG MOSFET based architecture, the RF input signal (($v_{RF} = (v_{rf} + V_{RF}) \sin(2\pi f_{RF} t)$) is applied at one gate while the local oscillator (LO) signal ($v_{LO} = (v_{lo} + V_{LO}) \sin(2\pi f_{LO} t)$) is applied at the another gate of the transistor. Here, v_{rf} and v_{lo} are the AC components of RF and LO signal respectively, while V_{RF} and V_{LO} are the respective DC bias components. The output signal ($V_{out} = A_v[\cos 2\pi t(f_{RF} - f_{LO}) - \cos 2\pi t(f_{RF} + f_{LO})]$) consisting of the two intermediate frequencies is observed at the drain of the DG MOSFET and A_v is $(v_{rf} + V_{RF})(v_{lo} + V_{LO})/2$. The conversion gain (CG) by definition, then becomes $(v_{lo} + V_{LO})/2$. However, this theoretical linear proportionality of CG on LO amplitude is not valid everywhere and there is a strong dependence on the device geometries and threshold as evident from this analysis, and this necessitates the requirement for bias optimization with quantum corrected simulations.

4.2.2. Non-linearity analysis

The DG-MOSFET Mixer's multiplicative/non-linear property has been analyzed here from Fig. 12a. The RF signal which is applied at the front gate is represented by small signal model. This is justified because the power level of RF signal is very small on reception at the antenna and remains small even amplified by the LNA. Therefore, the output voltage, V_{out} is given as:

$$V_{out} = g_m v_{RF} R_L \tag{5}$$

where g_m is the transconductance of the device at the front gate. The I-V characteristics of DG-MOSFET at saturation is modeled as [47]

$$I_D = K[(V_{gs} - V_T)^2 - K'e^{\frac{V_{gs} - V_0 - V_{ds}}{kT}}] \tag{6}$$

where K & K' are process and device constants and V_0 is a second order term of V_T [47]. Here the drain current at the front gate is modeled by ignoring the exponential term assuming a large V_{ds} at saturation, where the numerator at the exponent goes negative.

$$\frac{I_D}{2} \simeq K(V_{gs} - V_T)^2 \tag{7}$$

The transconductance at the front gate is,

$$g_m = \frac{1}{2} \frac{\partial I_D}{\partial V_{gs}} \tag{8}$$

From eqns. (7) and (8) we can write,

$$g_m = 2K(V_{gs} - V_T) \tag{9}$$

Now from eqns. (7) and (9),

$$g_m = \sqrt{(2KI_D)} \tag{10}$$

A large signal model is assumed for the back gate as the LO signal is locally generated and usually has high amplitude levels,

$$I_D/2 = K(v_{LO} - V_T)^2 \tag{11}$$

implies,

$$I_D = 2K(v_{LO} - V_T)^2 \tag{12}$$

Therefore from eqns. (5), (10) & (12),

$$V_{out} = K''(v_{LO} - V_T)v_{RF} \qquad (13)$$

Here K'' is a constant which include the process and device parameters and the resistor R_L. The eqn. (13) analyzes the DG-MOSFET device analysis for non-linear RF Mixer operation. The output voltage is the product of two input voltages. The process dependent parameter V_T can be eliminated if we consider a balanced/differential mixer mode. However, typically the balanced mode is avoided because in a receiver design the mixer follows the LNA which is generally single ended as it follows a single ended antenna. A balun which consumes a large area is thus required to construct before the mixer for the differential mode use.

4.2.3. Operating point analysis

After the FFT analysis (Fig. 13a inset) of the output at a very high temporal resolution, we observe significant spectral lines at the two intermediate frequencies of 40 MHz (f_{RF} - f_{LO}) and 60 MHz (f_{RF} + f_{LO}) indicating the appropriate double gate mixing performance and non-linearity. The presence of higher harmonics (such as at 100 MHz frequency) in the spectra indicates higher-order non-linearities and must be filtered out to work with the desired frequency. For the analysis purposes and simplification of the observed spectra, the LO signal used in our study is a pure sine-wave with a DC offset providing the operating point for the device, while the RF AC input at the another gate is held constant without a DC offset.

Our study indicates that the CG of the mixer rapidly changes with the amplitude of the LO rising to 200 mV (Fig. 13b), beyond which the increase is limited. Hence, for all L_{gate} values, the operating point of the mixer is chosen to be set around 120 mV for optimum power efficiency and CG. Similar results were also obtained for different N_As and t_{Si}s from corresponding analyses.

The CG is particularly sensitive to the LO DC bias (Fig. 13c) with an 'm-shape' dependence, where the middle dip could be as much as -80 dB. Hence, seemingly there are two bias conditions that provide similar performance in CG of the mixer (Fig. 13b). For instance, these two bias points are observed at 0.3 V and 1 V for L_{gate} = 30 nm and v_{lo} = 40 mV. Moreover, this m-shape is a very weak function of LO AC bias and L_{gate}. Data recorded with AC inputs of 40 mV with 120 mV shift mainly vertically with a large lateral similarity in terms of DC bias dependence. Likewise, the peak position shifts roughly 0.1 V only, as the gate length is varied from 90 nm to 30 nm. It is interesting to note that these optimum DC-bias ranges correspond to the least 'linear' sections of the device operation, as can be seen from the transfer characteristics and transconductance (g_m vs. I_d) curves in Fig. 13d. The current changes in a very non linear pattern around the optimum bias ranges and the g_m peak corresponds to the central dip in Fig. 13c. Clearly, the lower bias point ($\sim$ 0.3 V in Fig. 13c) should be preferred because of power efficiency and better stability indicated by the broader plateau. Similar analyses conducted for different N_A and t_{Si} of the DG-MOSFET mixer yield in similar results to our study of L_{gate}. A double-peaked LO-DC behavior persists in all cases. Summarizing results from these simulations, Tables 1, 2 and 3 list the optimum (lower) bias points for different structural parameters studied.

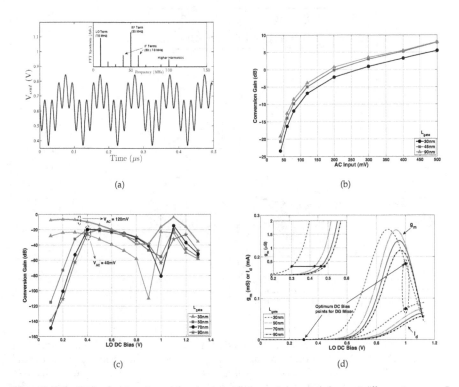

(a)   (b)

(c)   (d)

Figure 13. a) The FFT (inset) of the voltage at the mixer output (main panel) shows both the sum & difference terms as well as additional higher order harmonics. b) Variation of conversion gain with AC Input for different L_{gate}s. The CG increases rapidly before 120mV, after which the performance of the conversion gain is limited. c) Variation of CG with DC bias at different L_{gate}s. Observation of two AC inputs (120 mV & 40 mV) shows their CG variation with DC bias is similar. d) Transconductance (g_m) & drain current (I_d) over DC bias for different L_{gate}s. Out of two optimum bias points, the lower one at 0.3 V (30 nm) is chosen for better stability and power efficiency.

L_{gate} (nm)	30	50	70	90
DC Bias (V)	0.30	0.41	0.45	0.47

Table 1. Optimum LO DC bias for different gate lengths at $N_A = 10^{15}\ cm^{-3}$ & $t_{Si} = 5\ nm$

N_A (cm^{-3})	10^{15}	10^{16}	10^{17}	10^{18}
DC Bias (V)	0.45	0.47	0.48	0.50

Table 2. Optimum LO DC bias for different doping concentrations at $L_{gate} = 45\ nm$ & $t_{Si} = 5\ nm$

t_{Si} (nm)	5	10	20	30
DC Bias (V)	0.6	0.5	0.4	0.3

Table 3. Optimum LO DC bias for different body thicknesses at $N_A = 10^{15}\ cm^{-3}$ & $L_{gate} = 90\ nm$

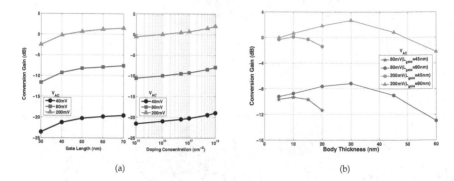

(a)   (b)

Figure 14. a) Dependence of CG on the gate length (L_{gate}) & doping concentration (N_A) for different AC Inputs. The weak correlation of these two parameters on the CG is clearly evident. b) Dependence of CG on the body thickness (t_{Si}) for different AC Inputs. CG varies with L_{gate} because of short channel effects.

4.2.4. Structural parameters

Next, we study the dependence of CGs recorded at the various LO AC amplitudes and at optimum DC (lower peak) bias conditions as a function of most significant structural parameters of the DG-MOSFET used for mixing. The results are summarized in Figs. 14a & 14b, which show the dependence of conversion gain with L_{gate}, N_A & t_{Si}. Clearly, the L_{gate} has almost no impact on the conversion gain at higher values while at lower value the impact becomes more pronounced. For N_A, the conversion gain almost remains constant at low doping levels whereas it slightly increases at very high (impractical) doping levels. However, from Fig. 14b we find that t_{Si} is a more significant parameter for conversion gain optimization. In a given gate length there appears to be an optimum body thickness that maximizes the CG. For example, at L_{gate} of 45 nm and 90 nm, the optimum body thickness is 10 nm and 30 nm, respectively. At either extreme of these values, the conversion gate is compromised due to the short channel effects in the higher end and quantum size effects at the lower end. Thus it is important to include both 2D/3D simulations and quantum corrections to optimize mixing performance in such nano-scale transistors, as with the case in this study.

We like to draw attention that the weak dependence of performance on the structural parameters here is a result of careful bias optimization. It also indicates that the choice of bias conditions, particularly the LO DC bias, is the most dominant handle in using DG-MOSFET active mixer. Admittedly, this observation may be counter intuitive, because the short channel effect are well known to adversely impact analog performance of the conventional MOSFETs in sub-100 nm regime. However, these adverse impacts are mostly related to the increase of the non-linearity in g_m which is certainly helpful for a mixer. In any case the well-scaled

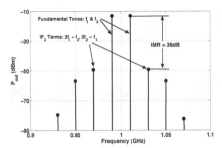

Figure 15. Computation of IP$_3$ using two tone frequency analysis (0.99 GHz and 1.01 GHz) around 1 GHz. The Intermodulation Ratio (IMR) gives level difference between the fundamental and the IP$_3$ terms and is used to obtain the IIP$_3$ = P$_{in}$(f$_1$,f$_2$) + IMR/2.

nature of the DG-MOSFET minimizes the emergence of strong short channel effects for the mixer performance.

Moreover, the apparent stability of mixer performance with device geometrical scaling could affect the phase noise in both positive and negative fashion. In terms of inter-device performance variations, the DG-MOSFETs will not suffer as much as the logic applications as the process variations in geometry does not appear to be a worry. However, since the LO-DC bias is the most important figure of merit, variations in threshold among devices and biasing errors/variations in circuits can be the main source of phase noise and limit the performance.

4.2.5. Linearity analysis

Finally, we examine the circuit for linearity implementing the two tone frequency analysis (Fig. 15). The 3rd order Input Intercept point (IIP$_3$) is found to be 15.9 dBm for 2 dBm LO power, indicating the suitability of the circuit [35].

4.3. Envelope detector

The demodulation of a non-coherent modulated wave requires an envelope detector. The envelope detector is basically a rectifier circuit that generates an envelope of the incoming high frequency carrier signal and strips off the carrier to recover the data.

In Fig. 16a, we have illustrated a 45 nm DG-MOSFET envelope detector circuit in which the output is inverted to that of binary input (Refer Fig. 6). The output signal needs further to be passed through an inverter for the recovery of the original signal. Although requires additional hardware, this circuit has an advantage over the straightforward recovery as the former has a better output swing over the latter [48]. The simulation (Fig. 16b) illustrates the recovered binary input information as same as that is shown in Fig. 6. The high frequency noise present with logic 1 data at the output can be easily filtered out.

4.4. Charge pump Phase Frequency Detector

The Phase Frequency Detector (PFD) is one of the two major components of a PLL, that is used for the demodulation of coherent modulated signal. The other being the local oscillator/VCO. It consists of two D Flip Flops and a reset circuit. The two D Flip Flops are implemented with eight NOR gates (four each) [49]. The reset path consists of another

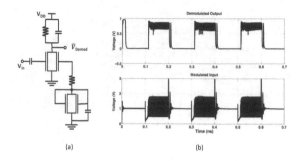

(a)   (b)

Figure 16. a) Envelope Detector Circuit with only two DG MOSFETs. b) The modulated input consisting of the both the carrier and data; the recovered demodulated output consisting only of data sans the carrier.

NOR gate. Here, each of the NOR gates are constructed with DG-MOSFETs. The circuit also consists of two DG-MOSFET NMOS switches implemented in regular V_T configuration (Fig. 17a).

4.4.1. DG-MOSFET NOR gate

The DG MOSFET NOR gate consists of only two DG transistors instead of four as in conventional CMOS architecture (Fig. 17b). This was first proposed by Chiang et. al [50]. The design employs the threshold-voltage (V_T) difference between double-gated and single-gated modes in a high V_T DG device to reduce the number of transistors by half.

The NOR logic with DG-MOSFETS is shown in Fig. 17c. One of the major advantages of using NOR gates using DG-MOSFETs is speed. The area and capacitance of the DG-MOSFET NOR gate is almost 2x less than the conventional CMOS due to reduced transistor count (half that of conventional CMOS) and associated isolation and wirings which lowers the capacitance and speeds up the circuit. In Fig. 17d we demonstrate this fact for different supply voltages. The advantage in higher speed is crucial for tiny phase error detection in PLL and is the subject of the following section.

4.4.2. Design and analysis

This analysis is carried with a supply of 1 V for $(W/L)_p$ = 4 μm / 45 nm and $(W/L)_n$ = 1 μm / 45 nm. The power consumed by the DG-MOSFET based Charge Pump PFD is 3.4 mW which is 21% less than that of the conventional CMOS under identical device dimensions and parameters. Although the drive current (I_{ON}) for DG-MOSFET is higher for both regular and high V_T configurations than that of a single gate MOSFET, the reduction in the number of transistor counts, reduces the power consumption decently. The area is also reduced almost by half resulting from this reduction.

The phase error between two pulses A and B can be correctly detected for both conventional CMOS and DG-MOSFET when the phase error between the two pulses (T_{pe}) is above a certain threshold. Now from our analysis at the previous section on speed enhancement of DG-MOSFET based NOR architecture we can deduce the rise time of the DG-MOSFET (T_{thDG}) is faster than that for rise time of conventional CMOS (T_{thSG}) to reach the desired threshold of logic 'HIGH' and thus we can write $T_{thDG} < T_{thSG}$. However, for low phase

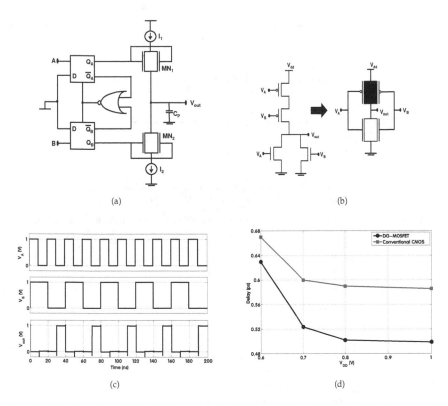

Figure 17. a) The Charge Pump PFD circuit implemented with DG-MOSFETs. b) 2-Input NOR Logic Gate in Conventional CMOS and its equivalent in DG-MOSFET. Two transistors are required in the DG-MOSFET. The PMOS in DG-MOSFET is kept at a high-V_T symbolized by a filled transistor. c) 2-Input DG-MOSFET NOR Logic simulated waveform for $V_{DD} = 1$ V. d) Delay comparison of 2-input NOR gate between conventional CMOS and DG-MOSFET for different supply voltages.

error applications, when $T_{thDG} \leq T_{pe} \leq T_{thSG}$, the PFD ceases to work correctly for the conventional CMOS. As observed from Fig. 18, for $T_{pe} = 80$ ps, the voltage at the output Q_A of the flip flop fails to reach the threshold to switch on the transistor MN_1 in the period when A is 'HIGH' and B is 'LOW'. The voltage *only* reaches the threshold when both A and B are HIGH. When B is high the voltage at Q_B also reaches 'HIGH' which turns the transistor MN_2 'ON'. Therefore, when both Q_A and Q_B are 'HIGH' (reaches the V_T) simultaneously, the current I_1 instead of charging the capacitor C_P passes through the switch MN_2. Thus the output voltage (V_{out}) remains nearly constant and changes only by a fraction of what should be in order to send the accurate message of phase error to the VCO, which follows the PFD in a PLL architecture. As a matter of fact, the V_{out} changes only by a meagre 0.005 mV for 100 ns. This is negligible and an incorrect feedback to the VCO. The V_{out} characteristics is verified from Fig. 19. This is the familiar dead zone condition where there is no or negligible charge pump current that contributes to no change in V_{out}.

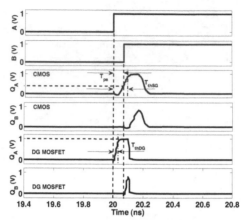

Figure 18. Phase error characteristics of two pulses A & B for conventional CMOS and DG-MOSFET for a phase error of 80 ps.

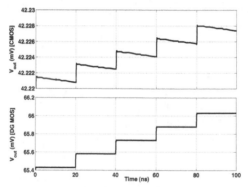

Figure 19. Charge Pump output voltage characteristics of Conventional CMOS and DG-MOSFET when $T_{thDG} \leq T_{pe} \leq T_{thSG}$

On the other hand, the advantage of DG-MOSFET is clearly evident from Fig. 18 where it can be confirmed that for the same period the threshold for the DG-MOSFET reaches the logic 'HIGH' when A is 'HIGH' and B is 'LOW'. Thus the current I_1 cannot escape through MN_2 and charges C_P instead. This is clearly because even when $T_{pe} \leq T_{thSG}$, the inequality $T_{pe} \geq T_{thDG}$, is still valid due to the fact that $T_{thDG} < T_{thSG}$ owing to the lower capacitance as discussed in the previous section. Thus the dead zone is avoided with the correct and significant change of 0.6 mV in V_{out} for the same duration as that of conventional CMOS (Fig. 19).

5. Summary and future prospects

The chapter has provided examples for unique and significant performance improvements available via a novel transistor architecture (FinFET or DG-MOSFET) in a wide collection of analog and mixed-signal circuits that can be used in today's integrated wireless communication, satellite navigation systems and sensor networks, verified through industry standard SPICE simulations. In particular, the chapter documents the tunable frequency response in relaxation and LC oscillators along with the gain tuning in wide-band PA and the LNA circuits. In all these cases, the performance improvements and tunable characteristics can be achieved via the availability of independently biased second gates in these new device architectures. In addition to gain tuning, the PA and the LNA performance parameters such as gain, bandwidth, linearity, NF are either comparable or better than some of the recent designs in conventional CMOS or III-V technologies. The fact that DG-MOSFET circuits utilize reduced transistor count compared to single gate CMOS is exemplified by relaxation oscillator, RF Mixer, OOK Modulator, Envelope Detector and Charge Pump PFD circuits. As obvious, the reduced transistor count aids in reducing area and may lower power dissipation. The biasing optimization technique of the RF Mixer described here maximizes the conversion gain of the RF Mixer with power efficiency. The DG-MOSFET Charge Pump PFD circuit avoids dead zone in PLL for low phase difference applications which is not possible in conventional CMOS as demonstrated here. The primary reason for this is reduced delay because of reduced area which in turn is achieved as a result of reduced transistor count.

With fabrication processes of DG-MOSFETs soon coming up with initiation from TSMC and rapidly expanding system-level efforts led by several national and international programs in US, Japan and Europe, along with several companies (such as Intel [4]) and academic centers focussing on these DG-MOSFET/FinFET/3DMOSFET technologies, we should expect a wide range of tunable analog RF circuits, reconfigurable logic blocks, on-chip power management blocks and mixed-signal system-on-chip applications to come into existence in the next few years.

Ultimately, with the ongoing nanotechnology revolution further performance improvements and architectural changes in devices are to be expected in the next decade and beyond. Our work here shows that such changes can be utilized by circuit engineering to result in very compact and capable systems, even when the actual change is to include merely an additional gate in the MOSFET architecture. This indicates that circuit engineering has a lot more to say not only in the final stretch of Moore's scaling, extending perhaps until 2020, but also in post-Moore area where fundamental fabric of building circuits may be altered significantly, and novel devices architectures and materials such as graphene, carbon nanotube, nanowire or molecular transistors are likely to play a significant role.

Acknowledgement

This research was partially supported by the NSF Award ECCS-1129010. We are thankful to the Co-PIs of this award, Prof. Avinash Kodi of Ohio University and Prof. David Matolak of University of South Carolina (formerly of Ohio University) for their support.

Author details

Soumyasanta Laha* and Savas Kaya

* Address all correspondence to: sl922608@ohio.edu

School of Electrical Engineering & Computer Science, Ohio University, Athens, OH, USA

References

[1] G. K. Celler and Sorin Cristoloveanu. Frontiers of silicon-on-insulator. *J. of Applied Physics*, 93:4956–4978, 2003.

[2] Thomas Skotnicki, James A. Hutchby, Tsu-Jae King, H.-S.Philip Wong, and Frederic Boeuf. The end of CMOS scaling. *IEEE Circuits Devices Mag.*, pages 16–26, 2005.

[3] J. Colinge. Multi-gate SOI MOSFETs. *Microelectronic Engineering (Elsevier)*, 84:2071–2076, 2007.

[4] K. Ahmed and K. Schuegraf. Transistor Wars. *IEEE Spectr.*, page 50, Nov 2011.

[5] Isabelle Ferain, Cynthia A. Colinge, and Jean-Pierre Colinge. Multigate transistors as the future of classical metal-oxide-semiconductor field-effect transistors. *Nature*, 479:310–316, Nov 2011.

[6] K. Roy, H. Mahmoodi, S. Mukhopadhyay, H. Ananthan, A. Bansal, and T. Cakici. Double-gate SOI Devices for Low-Power and High-Performance Applications. In *IEEE/ACM Int. Conf. on CAD*, pages 217 – 224, Nov 2005.

[7] Jae-Joon Kim and K. Roy. Double Gate-MOSFET Subthreshold Circuit for Ultralow Power Applications. *IEEE Trans. Electron Devices*, 51:1468–1474, 2004.

[8] A. Amara and O. Rozeau (Eds.). *Planar Double-Gate Transistor*. Springer, 2009.

[9] S. Kaya, H.F.A Hamed, and S. Laha. *Tunable Analog and Reconfigurable Digital Circuits with Nanoscale DG-MOSFETs*, chapter 9. INTECH, Rijeka, Croatia, Feb 2011.

[10] S. A. Tawfik and V. Kursun. Robust finFET memory circuits with P-Type data access Transistors for higher Integration Density and Reduced Leakage Power. *J. of Low Power Electronics*, 5:1–12, 2009.

[11] Meishoku Masahara et. al. Optimum Gate workfunction for Vth-Controllable Four-Terminal-driven double-gate MOSFETs (4t-XMOSFETs)-band-edge Workfunction versus Midgap Workfunction. *IEEE Trans. Nanotechnol.*, 5:716–722, Nov 2006.

[12] Hanpei Koike and Toshihiro Sekigawa. XDXMOS: A novel technique for the Double-Gate MOSFETs Logic Circuits. In *IEEE Custom Integrated Circuits Conference (CICC)*, pages 247–250, 2005.

[13] Kazuhiko Endo et. al. Four-Terminal FinFETs fabricated using an etch-back gate separation. *IEEE Trans. Nanotechnol.*, 6:201–205, Mar 2007.

[14] D. Wilson, R. Hayhurst, A. Oblea, S. Parke, and D. Hackler. Flexfet: Independently-double-gated SOI Transistor with Variable vt and 0.5v operation achieving near ideal subthreshold slope. In *IEEE Intl. SOI Conf.*, pages 147–148, 2007.

[15] K. Modzelewski, R. Chintala, H. Moolamalla, S. Parke, and D.Hackler. Design of a 32nm Independently-Double-Gated flexFET SOI Transistor. In *IEEE University/Government/Industry Micro/Nano Symposium (UGIM)*, pages 64 – 67, 2008.

[16] J.P Raskin, T.M. Chung, V. Kilchytska, D. Lederer, and D. Flandre. Analog/rf performance of multiple gate SOI devices: wideband simulation and characterization. *IEEE Trans. Electron Devices*, 53:1088–1095, May 2006.

[17] A. Lazaro and B. Iniguez. RF and Noise performance of Double Gate and Single Gate SOI. *Solid State Elect.*, 50:826–842, 2006.

[18] B. IÃśiguez, A LÃązaro, O. Moldovan, A. Cerdeira, and T. A.Fjeldly. DC to RF Small-Signal Compact DG MOSFET model. In *NSTI-Nanotech*, pages 680–685, 2006.

[19] Gen Pei and Edwin Chih-Chuan Kan. Independently driven DG MOSFETs for Mixed-Signal Circuits: Part I - Quasi-Static and Nonquasi-Static Channel Coupling. *IEEE Trans. Electron Devices*, 51:2086–2093, Dec 2004.

[20] Xiaoping Liang and Yuan Taur. A 2-d Analytical Solution for SCEs in DG MOSFETs. *IEEE Trans. Electron Devices*, 51:1385–1391, Aug 2004.

[21] Huaxin Lu and Yuan Taur. An analytic Potential Model for Symmetric and Asymmetric DG MOSFETs. *IEEE Trans. Electron Devices*, 53:1161–1168, May 2006.

[22] Marina Reyboz, Olivier Rozeau, Thierry Poiroux, Patrick Martin, and Jalal Jomaah. An explicit analytical charge-based model of undoped independent double gate MOSFET. *Solid State Elctronics (Elsevier)*, 50:1276–1282, 2006.

[23] M. Reyboz, P. Martin, T. Poiroux, and O. Rozeau. Continuous model for independent double gate mosfet. *Solid State Elctronics (Elsevier)*, 53:504–513, 2009.

[24] Wei Zhao and Yu Cao. Predictive Technology Model for Nano-CMOS Design Exploration. *ACM Journal on Emerging Technologies in Computing Systems*, 3:1–17, 2007.

[25] J. G. Fossum. *UFDG Users Guide (Ver. 3.0)*. University of Florida Press, Gainesville, FL, 2004.

[26] L. Ge, J.G. Fossum, and B. Liu. Physical Compact Modelling and Analysis of velocity overshoot in Extremely scaled CMOS Devices and Circuits. *IEEE Trans. Electron Devices*, 48:2074–2080, Sep 2001.

[27] K. Kim. *Design and analysis of Double gate CMOS for Low Voltage Integrated Circuit Applications, including physical modeling of Silicon-on-Insulator MOSFETs*. University of Florida Dissertation, Gainesville, FL, 2001.

[28] S. Laha, K. Wijesundara, A. Kulkarni, and S. Kaya. Ultra-compact low-power ICO/VCO circuits with double gate MOSFETs. In *IEEE International Semiconductor Device Research Symposium (ISDRS)*, pages 1–2, 2011.

[29] D. Jimenez, B. Iniguez, J. Sune, and J.J. Saenz. Analog Performance of the Nanoscale Double-Gate Metal-Oxide-Semiconductor field-effect-transistor near the ultimate scaling limits. *Journal of Applied Physics*, 96:5271 – 5276, Nov 2004.

[30] Behzad Razavi. *RF Microelectronics*. Prentice Hall, Boston, USA, 2012.

[31] F. Xiong. *Digital Modulation Techniques*. Artech House, Boston, 2006.

[32] A. Hajimiri and T.H. Lee. A general theory of phase noise in electrical oscillators. *IEEE J. Solid-State Circuits*, 33:179–194, Feb 1998.

[33] Guansheng Li and Ehsan Afshari. A Low-Phase-Noise Multi-Phase Oscillator Based on Left-Handed LC-Ring. *IEEE J. Solid-State Circuits*, 45:1822–1833, Sep 2010.

[34] S. Laha, S. Kaya, A. Kodi, and D. Matolak. 60 GHz OOK Transmitter in 32 nm DG FinFET Technology. In *IEEE International Conference on Wireless Information Technology and Systems (ICWITS) (Accepted to appear)*, 2012.

[35] T.H. Lee. *The Design of CMOS Radio-Frequency Integrated Circuits*. Cambridge University Press, Cambridge, UK, 2004.

[36] S. Laha, S. Kaya, A. Kodi, and D. Matolak. Double Gate MOSFET based Efficient Wideband Tunable Power Amplifiers. In *IEEE Wireless and Microwave Conference (WAMICON)*, pages 1–4, 2012.

[37] Pin-Cheng Huang, Kun-You Lin, and Huei Wang. A 4-17 GHz Darlington Cascode Broadband Medium Power Amplifier in 0.18 micron CMOS Technology. *IEEE Microw. Wireless Compon. Lett.*, 20:43–45, Jan 2010.

[38] Terry Yao, Michael Gordon, Kenneth Yau, M.T. Yang, and Sorin P. Voinigescu. 60-GHz PA and LNA in 90-nm RF-CMOS. In *IEEE Radio Frequency Integrated Circuits (RFIC) Symposium*, pages 4–7, 2006.

[39] Wong-Sun Kim, Xiaopeng Li, and M. Ismail. A 2.4 GHz CMOS low noise amplifier using an inter-stage matching inductor. In *IEEE Midwest Symposium on Circuits and Systems*, pages 1040 – 1043, 2000.

[40] Michael Gordon, Terry Yao, and Sorin P. Voinigescu. 65-GHz Receiver in SiGe BiCMOS using Monolithic Inductors and Transformers. In *IEEE Topical Meeting on Silicon Monolithic Integrated Circuits on RF Systems Digest*, pages 1–4, 2006.

[41] Chi-Chen Chen, Yo-Sheng Lin, Pen-Li Huang, Jin-Fa Chang, and Shey-Shi Lu. A 4.9-dB NF 53.5-62 GHz micro-machined CMOS Wideband LNA with Small Group-Delay-Variation. In *IEEE International Microwave Symposium Digest*, pages 489 – 492, 2010.

[42] S. Pellarano, Y. Palaskas, and K. Soumyanath. A 64 GHz LNA with 15.5 dB Gain and 6.5 dB NF in 90 nm CMOS. *IEEE J. Solid-State Circuits*, 43:1542–1552, Jul 2008.

[43] Shuai Huang, Xinnan Lin, Yiqun Wei, and Jin He. Derivative Superposition method for DG MOSFET application to RF mixer. In *Asia Symposium on Quality Electronic Design (ASQED)*, page 361, 2010.

[44] M. V. R Reddy, D. K. Sharma, M. B. Patil, and V. R. Rao. Power-Area Evaluation of Various Double-Gate RF Mixer Topologies. *IEEE Electron Device Lett.*, 26:664–666, 2005.

[45] W. Zhang, J.G. Fossum, L. Mathew, and Y. Du. Physical Insights Regarding Design and Performance of Independent-Gate FinFETs. *IEEE Trans. Electron Devices*, 52:2198–2206, 2005.

[46] S. Laha, M. Lorek, and S. Kaya. Optimum Biasing and Design of High Performance Double Gate MOSFET RF Mixers. In *IEEE International Symposium on Circuits and Systems (ISCAS)*, pages 3278–3281, 2012.

[47] Y. Taur et al. A a continuous, Analytic Drain-Current model for DG MOSFETs. *IEEE Electron Device Lett.*, 25:107–109, Feb 2004.

[48] Jri Lee, Yentso Chen, and Yenlin Huang. A Low-Power Low-Cost Fully-Integrated 60-GHz Transceiver System with OOK Modulation and On-Board Antenna Assembly. *IEEE J. Solid-State Circuits*, 45:264–275, Feb 2010.

[49] B. Razavi. *Design of Analog CMOS Integrated Circuits*. McGraw Hill, Boston, USA, 2001.

[50] M.H. Chiang, K. Kim, C.T. Chuang, and C. Tretz. High-density reduced-stack logic circuit techniques using independent-gate controlled double-gate devices. *IEEE Trans. Electron Devices*, 53:2370–2377, Sep 2006.

Chapter 2

A Successive Approximation ADC using PWM Technique for Bio-Medical Applications

Tales Cleber Pimenta, Gustavo Della Colletta,
Odilon Dutra, Paulo C. Crepaldi,
Leonardo B. Zocal and Luis Henrique de C. Ferreira

Additional information is available at the end of the chapter

1. Introduction

Analog to digital (A/D) converters provide the interface between the real world (analog) and the digital processingdomain. The analog signals to be converted may originate from many transducers that convert physical phenomena like temperature, pressure or position to electrical signals. Since these electrical signals are analog voltage or current proportionals to the measured physical phenomena, its necessary to convert them to digital domain to conduct any computational. Nowadays, the development of the IC technology resulted in a growth of digital systems. A/D converters are present in the automotive industry, embedded systems and medicine for example. Thus, A/D converters have become important and the large variety of applications implies different types of A/D conversions.

For the A/D type considerations, the analog input should be characterized as one of the following three basic signal types [3].

- Direct current (DC) or slowly varying analog signals.

- Continuous changing and single event alternating current (AC) signals.

- Pulse-amplitude signal.

For sampling the first type of signals, typical A/D conversion architectures are slope, voltage to frequency, counter ramp and sigma-delta. The second signal type is better sampled using the successive approximation, multistep and full parallel A/D conversion architectures. The last signal type uses successive approximation, multistep, pipeline and full parallel architectures.

After choosing the A/D converter architecture, it is important to keep in mind that any of them have nonlinearities that degrade the converter performance. These nonlinearities are accuracy parameters that can be defined in terms of Differential Nonlinearity (DNL) and Integral Nonlinearity (INL). Both have negative influence in the converter Effective Number of Bits (ENOB) [2].

- Differential Nonlinearity (DNL) is a measure of how uniform the transfer function step sizes are. Each one is compared to the ideal step size and the difference in magnitude is the DNL.

- Integral Nonlinearity (INL) is the code midpoints deviation from their ideal locations.

Therefore it is important to design implementations capable of improving the ADCs performance by improving DNL and INL.

Physiological signals have amplitudes ranging from tens of μV to tens of mV and the frequencies spanning from DC to a few KHz. By considering those features and the application requirements, in order to make a reliable conversion, A/D converter may not have missing codes and must be monotonic. This can be accomplished assuring that the DNL error is less then 0.5 of last significant bits (LSBs).

2. Biomedical Application

Advances in low power circuit designs and CMOS technologies have supported the research and development of biomedical devices that can be implanted in the patient. These devices have a sensor interface specially designed to acquire physiological signals, usually composed of an operational amplifier with programmable gain and reconfigurable band-width features, low pass filter and an A/D converter [8, 10]. The signals are acquired and digitalized in the sensor, thus protecting data from external noise interference.

Specific research on A/D converters for biomedical application is focused on design low power circuits regardless of the monotonic feature, once DNL error is above $0.5\ LSBs$, affecting the converter accuracy [5, 6]. The proposed Successive Approximation architecture offers both low power consumption and high accuracy features for use in biomedical applications.

3. Conventional SAR architectures

Figure 1 illustrates the block diagram of the conventional SAR architecture. It is composed of a Successive Approximation Register that controls the operation and stores the output converted digital data, of a digital-to-analog converter stage (DAC), a comparator usually built with a operational amplifier and of a sample and hold circuit. The output can be taken serially from the comparator output or parallel from the SAR outputs.

The operation consists on evaluating and determining the bits of the converted digital word, one by one, initiating from the most significant bit. Thus the SAR architecture uses n clock cycles to convert a digital word of n bits. The successive approximation architecture provides intermediate sample rates at moderate power consumption that makes it suitable for low power applications.

The internal DAC stage, illustrated in Figure 1 is usually designed using capacitor networks that are susceptible to mismatches caused by the fabrication process variation, since the design is based on absolute capacitance values. These mismatches affect the converter accuracy, thus increasing the DNL and INL errors.

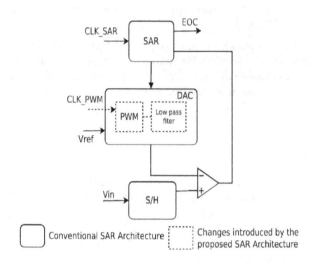

(a) Conventional and proposed architecture block diagrams.

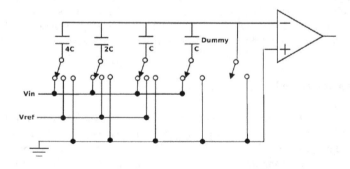

(b) Conventional internal DAC stage.

Figure 1. Conventional and proposed SAR architecture and conventional internal DAC stage.

4. Proposed Architecture

The presented architecture aims to eliminate the mismatches introduced during fabrication process by replacing the conventional internal DAC based on capacitor networks by a digital PWM modulator circuit and a first order low pass filter.

Figure 1 shows the block diagram of the proposed architecture (dotted line) as a modification on a conventional one (full line).

A PWM signal can be stated in terms of an even function, as illustrated in Figure 2 [1]. By using Fourier series, it can be represented in terms of equations (1) to (4).

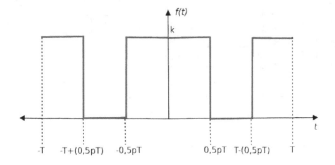

Figure 2. PWM signal stated as an even function.

$$f(t) = A_0 + \sum_{n=1}^{\infty} [A_n cos(\tfrac{2n\pi t}{T}) + B_n sin(\tfrac{2n\pi t}{T})] \qquad (1)$$

$$A_0 = \tfrac{1}{2T} \int_{-T}^{T} f(t) dt \qquad (2)$$

$$A_n = \tfrac{1}{2T} \int_{-T}^{T} f(t) cos(\tfrac{2n\pi t}{T}) dt \qquad (3)$$

$$B_n = \tfrac{1}{2T} \int_{-T}^{T} f(t) sin(\tfrac{2n\pi t}{T}) dt \qquad (4)$$

where A_0 represents the fundamental frequency, A_n states the even harmonics and B_n states the odd harmonics.

By performing the integral on a PWM signal with amplitude $(f(t)=k)$, the results are given by equations (5) to (7).

$$A_0 = kp \qquad (5)$$

$$A_n = k\frac{1}{n\pi}[sin(n\pi p) - sin(2n\pi(1 - \frac{p}{2}))]$$ (6)

$$B_n = 0$$ (7)

where p denotes the duty cycle.

That result shows that the PWM signal consists of a DC level and a square wave of zero average, as illustrated in Figure 3. Only the DC level is necessary in order to implement an internal DAC stage, since any DC level varying from zero to k can be obtained by selecting the proper duty cycle.

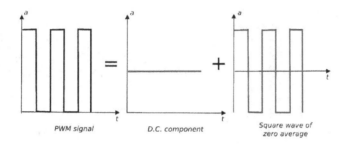

Figure 3. PWM signal split in a D.C level plus a square wave.

A way of recovering the DC level is to low pass filter the PWM signal. Since there is no ideal filter, the recovered DC level will have a certain ripple, as illustrated in Figure 4.

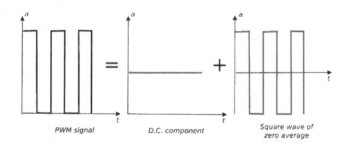

Figure 4. Low pass filtering the PWM signal.

4.1 Modeling

This section provides the modeling of a *4 bit* A/D Converter. Functional models for the SAR, PWM generator, Low pass filter and comparator blocks are discussed. Also the equating necessary to determine the filter features and clock frequencies is developed. SAR and PWM generator digital circuits are modeled using VHDL hardware description language. Comparator and the first order low pass filter are modeled using compartmental blocks.

A macro level simulation is performed using MatLab in order to validate the architecture. Electrical and post layout simulations are performed using Spectre simulator. The A/D converter Layout is developed in *0.5 μm* standard CMOS process using Cadence Virtuoso and NCSU Design Kit (Free design kit available from North Caroline State University).

4.1.1 Successive Approximation Digital Logic

The Successive Approximation logic evaluates every digital word output bit according to the clock (CLK) signal. Thus, initiating by the most significant bit, one by one, the bits are evaluated and determined, until the last significant bit. Figure 4 illustrates the SAR digital circuit. The control logic is based on a simple shift register. There is also a flip-flop array that stores the input selection (SEL) that is attached to the comparator output.

On a reset (RST) signal, the shift register is loaded with 10000 and the flip-flop array is loaded with 0000. The combinational logic based on OR gates assures the value 1000 at the output (Q_3-Q_0). When the first clock pulse arrives, the shift register value is changed to 01000 while the flip-flop array remains with the same value, except for the most significant bit, since it has been already determined. Thus, the SAR output will show something like $X000$, where X represents the previously determined value.

One special feature is to use an extra flip-flop in the shift register to indicate the end of conversion (END), enabling the converted digital word to be read in the rising edge of the fifth clock pulse.

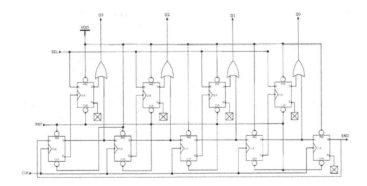

Figure 5. Successive Approximation Register.

4.1.2 Low Pass Filter

Circuits powered by *2.5V* using a *0.5 μm* standard CMOS process, as in this case, can operate at *2MHz* maximum frequency, limiting the operation to about *200 Hz* of sampling rate, regarding the proposed architecture design. These feature lead to a high value of capacitance in the RC first order low pass filter, which is impracticable to be integrated. An alternative used to validate the proposed architecture is the implementation of an external first order RC low pass filter, as show in Figure 6.

4.1.3 Digital PWM Modulator

The digital PWM modulator circuit is capable of varying the duty cycle of the output (PWM) according to the digital input word (D_3™ D_0). The circuit is illustrated in Figure 7 and consists of registers, a synchronous *4-bit* counter, a combinational reset and a combinational comparison logic.

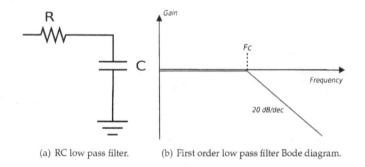

(a) RC low pass filter.   (b) First order low pass filter Bode diagram.

Figure 6. External RC first order low pass filter.

On a reset (RST) pulse, the counter resets to *0000* and the registers store the input word. The counter is incremented at every clock (CLK) cycle and the comparison logic assures that the output remains set while the counter does not reach the value stored into the registers. When it occurs, the output resets and the count continues until the counter reaches the end of counting. The reset logic makes the output flip-flop to set every time the counter resets, thus assuring that the output is set at the beginning of the counting. At this time, the registers are updated with the value present in the input (D_3- D_0) from the SAR output. The reset logic also has a flip-flop responsible for synchronizing the output of the AND gate to the clock signal, since the AND inputs arrive at different timings.

4.1.4 Inverter Based Comparator

The inverter based comparator circuit is used in order to decrease power consumption, since there is no quiescent power consumption. Figure 8 illustrates the comparator stage that uses a low power consumption architecture [7].

The circuit uses lagged clock signals to avoid overlapping, therefore assuring that the switches S_1, S_2 and S_3 do not close at the same time. At time ϕ_1, the switch S_2 is open and the switches S_1 and S_3 are closed, thus charging the capacitor C with V_{in}-V_{th}, where V_{th} is the inverter threshold voltage. Consequently any voltage variation during time ϕ_2 will be sensed by the inverter.

At time ϕ_2, the switches S_1 and S_3 are open and S_2 is closed, thus applying to the capacitor C the voltage produced by the PWM generator. This produces a voltage variation in the inverter input and the comparator makes the decision.

The switches S_1, S_2 and S_3 were replaced by solid state switches based on a nMOS transistor. After passing through a booster circuit, the clock signal is applied to the transistors gates.

4.1.5 Equating

The previous subsections illustrated the functional models for each stage of the proposed 4-bit A/D converter. Nevertheless is still necessary to determine the low pass filter features and the clock frequency for the digital stages, SAR, comparator and PWM generator.

The comparator must evaluate every time the SAR tests a new bit, so they have to be synchronized by the same clock signal. Assuming that all N bits must have to be determined before a new sampling begins, equation (8) states the clock frequency for the comparator and the SAR stage.

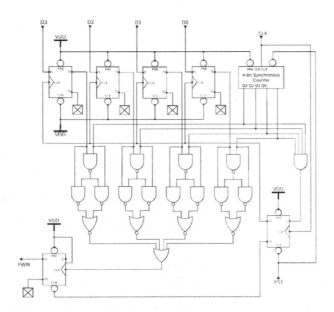

Figure 7. Digital Pulse Width Modulation generator.

$$f_{SAR} \geq f_s \times N \qquad (8)$$

where N represents the shift register number of bits, including the EOC bit and f_s represents the sampling rate.

Now, the low pass filter time constant ought to be determined. Equation (9) shows the cut off frequency for the first order filter.

$$f_c = \frac{1}{2\pi\tau} \tag{9}$$

where f_c represents the cut of frequency and τ states the filter time constant.

Assuming 5τ to accommodate a signal, equation (9) can be rewritten as equation (10)

$$f_c = \frac{1}{2\pi 5\tau} \tag{10}$$

From Figure 1, it can be observed that the filter must respond faster or at least at the same rate the SAR tests each bit. Thus, equation (11) states the maximum time constant for the low pass filter.

$$\tau \leq \frac{1}{2\pi 5 f_{SAR}} \tag{11}$$

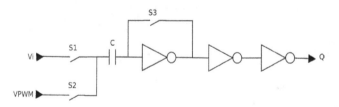

Figure 8. Inverter comparator circuit.

The frequency of the PWM signal must have to be characterized in order to be properly filtered. Since there is no ideal filter, the filtered signal will present a ripple. The PWM signal can be stated in terms of DC level and a sum of even harmonics, as in 12.

$$F_{PWM}(t) = A_0 + \sum_{n=1}^{\infty} A_n \cos\left(\frac{2n\pi t}{T}\right) \tag{12}$$

Taking into account only the even harmonics, as stated in 13, the energy carried by them can be determined.

$$g_n(t) = A_n \cos\left(\frac{2n\pi t}{T}\right), \ n = (0, 1, 2, \dots) \tag{13}$$

It is known that the energy is proportional to $\left(g_n^2(t)\right)$. The maximum energy occurs at $\frac{\partial}{\partial p} g_n^2(t) = 0$. Thus:

$$
\begin{aligned}
\frac{\partial}{\partial p} g_n^2(t) &= \frac{\partial}{\partial p}\left(A_n^2 cos^2(\frac{2n\pi t}{T})\right) \\
&= cos^2(\frac{2n\pi t}{T})\frac{\partial}{\partial p}(A_n^2) \\
&= cos^2(\frac{2n\pi t}{T})2A_n\frac{\partial}{\partial p}(A_n)=0
\end{aligned}
\tag{14}
$$

Equation 14 shows that the cosine term is independent of the duty cycle p and that the maximum energy occurs when $\frac{\partial}{\partial p} A_n = 0$, as shown in 15.

$$
\begin{aligned}
\frac{\partial}{\partial p} A_n &= \frac{\partial}{\partial p}(\frac{1}{n\pi}[sin(n\pi p)-sin(2n\pi(1-\frac{p}{2})]) \\
&= cos(n\pi p) + cos(2n\pi(1-\frac{p}{2})) \\
&= cos(n\pi p) + cos(2n\pi - n\pi p) \\
&= cos(n\pi p) + cos(2n\pi) \cdot cos(n\pi p) + sin(2n\pi) \cdot sin(n\pi p)=0
\end{aligned}
\tag{15}
$$

It can be observed that $cos(2\,n\,\pi)$ is unity for any value of n, the term $sin(2\,n\,\pi)$ is zero for any value of n. Thus, equation 15 can be rewritten in terms as 16.

$$
\frac{\partial}{\partial p} A_n = 2cos(n\pi p)=0
\tag{16}
$$

Equation 16 shows that the maximum energy in each harmonic is obtained at different duty cycles.

Since there is no ideal filter, after the low pass filtering, the harmonics will not be completely eliminated, but attenuated. It is necessary to evaluate the minimum attenuation required by system, once it is directly linked to ripple amplitude present in the filtered DC level.

Since the first harmonic caries the most energy, it is reasonable to take just it into account to characterize the low pass filter.

Thus, considering the first harmonic ($n=1$) and the maximum energy scenario $\left(p=\frac{1}{2}\right)$, isolating the first harmonic term $A_n cos(\frac{2n\pi t}{T})$, the maximum ripple expression can be expressed by 17. Figure 9 illustrates the PWM signal, where h_1 represents the ripple amplitude variation given by the first harmonic.

$$
h_1 = \frac{2k}{\pi}cos(\frac{2n\pi t}{T})
\tag{17}
$$

It is important to notice that the cosine term introduces a variation interval of $-\dfrac{2k}{\pi} \leq \dfrac{2k}{\pi}$ in the ripple amplitude. Equation 18 shows the maximum peak to peak variation.

$$h_{1_{pp}} = \frac{2k}{\pi} - \left(-\frac{2k}{\pi}\right) = \frac{4k}{\pi} \tag{18}$$

Figure 9 illustrates two sequential quantization levels defined by the filtered PWM signal. If the ripple present in two sequential quantization levels overlaps, the converter will lead to a wrong conversion.

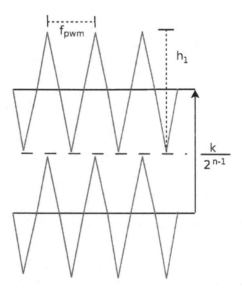

Figure 9. Maximum ripple amplitude.

Thus, equation (19) states the minimum attenuation necessary to keep ripple under an acceptable value.

$$-h_{1_{pp}} A \leq \frac{k}{2^{N-1}}$$

$$-\frac{4k}{\pi} A \leq \frac{k}{2^{N-1}}$$

$$A \geq \frac{\pi}{2^{N+1}} \tag{19}$$

$$A_{dB} \geq 20 log\left(\frac{\pi}{2^{N+1}}\right)$$

Since equation (19) expresses the attenuation in dB, the easier way to determine the PWM frequency is to plot the Bode diagram of the previously designed low pass filter and look directly into the frequency that provides the minimum necessary attenuation, as shown in Figure 10. Higher attenuation will decrease the ripple amplitude assuring the correct behavior of the A/D converter and a maximum attenuation is limited by the maximum frequency achieved by the PWM signal.

Finally, the PWM generator design requires a clock frequency 2^{N-1} times greater then output PWM signal, as stated by equation (20).

$$f_{pwm}^{clk} = 2^{N-1} f_{pwm} \qquad (20)$$

where f_{pwm}^{clk} states the clock frequency and f_{pwm} states the PWM signal frequency.

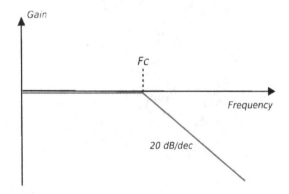

Figure 10. Determining the PWM signal frequency.

5. Simulations

The 4 ™ *bit* SAR ADC using PWM technique was designed for the ON *0.5 μ m* CMOS process using Cadence Virtuoso. simulations were conducted on Spectre simulator.

Figure 11 shows the circuit layout that occupies *0.749 mm²*. The main simulation results are given in table I.

It can be observed that the proposed architecture improved the A/D Converter accuracy, since the DNL and INL values are less then 0.1 LSB and also that it consumes low power.

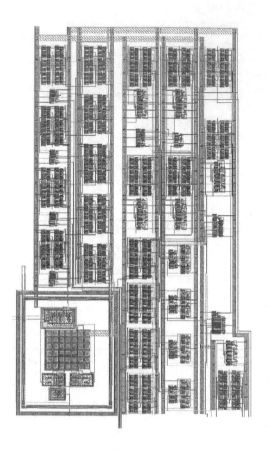

Figure 11. Circuit layout.

Technology	0.5 um
Supply Voltage	2.5 V
Max. Sampling frequency	200 Hz
ENOB (@166.67 Hz)	3.7549-b
DNL(max)	0.086 LSB
INL(max)	0.99 LSB
Power Consumption	16 uW
FoM (Figure of Merit)	7.11 nJ/conv.-step

Table 1. SAR ADC simulated performance.

Figure 12 shows the post layout simulation of DNL and INL for a slow ramp input. The values are good, lower than *0.086 LSB* and *0.1 LSB*, respectively, showing that the characteristic of proposed architecture does not differ too much form the ideal one.

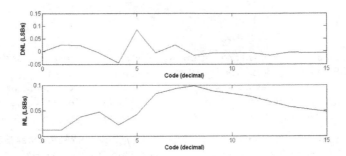

Figure 12. DNL and INL post simulation results.

Figure 13 illustrates the output frequency spectrum for a 32 point DFT. When ADC is tested with sinusoidal input at 166.67 Hz for a 15.63 Hz signal, it gives a good SNDR value of 24.36 dB, which results in 3.7549 effective number of bits (ENOB), thus proving the high accuracy achieved by the proposed architecture.

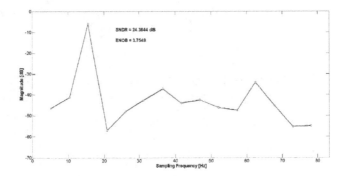

Figure 13. ADC simulated output frequency spectrum.

6. Future Research

The 4-bit layout was fabricated trough MOSIS education program. The prototypes will be tested and the results will be compared to the simulations.

After chip characterization, a proper integrated low pass filter will be implemented in a new prototyping. A new ADC with a larger number of bits will be developed in order to better investigate the non-linearities, ENOB and FoM results.

7. Conclusion

In order to validate the proposed architecture, a 4 ™ *bit* SAR A/D converter was designed in 0.5μ *m* CMOS standard process. The layout was developed using CADENCE Virtuoso and occupies *0.749 mm²*. Post-layout simulations conducted in Spectre simulator using the BSIM3v3 model show that the modifications introduced in the internal DAC stage contributed to minimize DNL *(0.086 LSB)* and INL *(0.099)* errors, as expected.

They also contributed to improve A/D converter accuracy, since the SNDR was improved to *24.36 dB* of *25.84 dB* maximum theoretical value, leading to *3.75* effective bits.

The feature of being almost fully digital contributes to reduce the circuit complexity, the silicon area and power consumption.

The features of high accuracy and low power consumption make the proposed architecture suitable for biomedical applications.

This architecture can be extended to build higher resolution converters by only adding more hardware to the digital stages or building pipeline structures.

Author details

Tales Cleber Pimenta*, Gustavo Della Colletta, Odilon Dutra, Paulo C. Crepaldi, Leonardo B. Zocal and Luis Henrique de C. Ferreira

*Address all correspondence to: tales@unifei.edu.br

Universidade Federal de Itajuba-UNIFEI, Brazil

References

[1] Alter, D. M. (2008). Using pwm output as a digital-to-analog converter on a tms320f280x digital signal controller. Technical report, Texas Instruments.

[2] Eid, E.-S., & El-Dib, H. (2009). Design of an 8-bit pipelined adc with lower than 0.5 lsb dnl and inl without calibration. *Design and Test Workshop (IDT), 2009 4th International*, 1-6.

[3] Hoeschele, D. F. J. (1994). Analog-to-Digital and Digital-to-Analog Conversion Techniques. John Wiley e Sons, 2nd edition.

[4] Lin, Y. Z., Liu, C.C., Huang, G. Y., Shyu, Y. T., & Chang, S. J. (2010). A 9-bit 150-ms/s 1.53- mw subranged sar adc in 90-nm cmos. *VLSI Circuits (VLSIC), 2010 IEEE Symposium on*, 243-244.

[5] Lu, T. C., Van, L. D., Lin, C. S., & Huang, C.-M. (2011). A 0.5v 1ks/s 2.5nw 8.52-enob 6.8fj/conversion-step sar adc for biomedical applications. *Custom Integrated Circuits Conference (CICC), 2011 IEEE*, 1-4.

[6] Mesgarani, A., & Ay, S. (2011). A low voltage, energy efficient supply boosted sar adc for biomedical applications. *Biomedical Circuits and Systems Conference (BioCAS), 2011 IEEE*, 401-404.

[7] Mikkola, E., Vermeire, B., Barnaby, H., Parks, H., & Borhani, K. (2004). Set tolerant cmos comparator. *Nuclear Science, IEEE Transactions on*, 51(6), 3609-3614.

[8] Ng, K., & Chan, P. (2005). A cmos analog front-end ic for portable eeg/ecg monitoring applications. *Circuits and Systems I: Regular Papers, IEEE Transactions on*, 52(11), 2335-2347.

[9] Talekar, S., Ramasamy, S., Lakshminarayanan, G., & Venkataramani, B. (2009). A low power 700msps 4bit time interleaved sar adc in 0.18um cmos. *TENCON 2009-2009 IEEE Region 10 Conference*, 1-5.

[10] Zou, X., Xu, X., Yao, L., & Lian, Y. (2009). A 1-v 450-nw fully integrated programmable biomedical sensor interface chip. *Solid-State Circuits, IEEE Journal of*, 44(4), 1067-1077.

Analog CAD

Chapter 3

Interval Methods for Analog Circuits

Zygmunt Garczarczyk

Additional information is available at the end of the chapter

1. Introduction

Concepts of interval analysis are suitable for solving some problems for linear and nonlinear analog circuits. The effects of circuit parameter uncertainties on circuit performance are always of great concern to the system designer. It is desirable to know a priori estimates of the circuit response, subject to parameter uncertainties. In the first section of this chapter a problem of calculating of the operating regions (solutions) for linear circuits with parameters done as interval numbers is considered. Such circuits are described by linear interval equations. The set of all possible operating points of the circuit may have a very complicated structure and it is usually impractical to calculate it. Applying ideas of interval analysis it is possible to calculate multidimensional rectangular region bounding the set of operating points. In the paper an algorithm of iterative evaluation of the bounds of operating regions is applied. The second section deals with the finding DC solutions of nonlinear, inertialess circuits. Using idea of a continuation method one can solve nonlinear equations describing circuit by tracing so-called solution path. A version of the predictor-corrector method for computing points of continuation path of a nonlinear equation is presented. The target of this approach is to control the corrector step in such a manner that the predictor step is sufficiently large but that the corrector step does not jump to another continuation path. Additionally interval analysis offers possibility to solve that problem by applying generalized bisection method based on some interval operator associated with nonlinear equation. In the section Krawczyk operator is used in n-dimensional box-searching of all solutions. For both sections numerical studies are reported in order to illustrate and verify presented approach.

2. Analysis of linear circuits with interval data

The changes in the performance of a linear circuit due to the variations in circuit parameters are of great practical importance in engineering analysis and design. Such perturbations of parameters may model the effects of uncertainties in manufacturing tolerances, ageing of components, environmental causes and the like. because of these uncertainties, the value of the parameters of a given linear circuit may frequently be treated as belonging to suitable intervals.

From this point of view, the aim of this paper is to consider solutions of linear circuits using paradigm of interval computations [1]. This problem is closely related to the tolerance analysis of electric and electronic circuits especially the worst-case analysis [3].

To explain problem formulation let us consider following simple example. We are interested in computation of nodal voltages in the R-ladder circuit of Fig.1.

Parameters of the circuit are done as interval number, viz. $E \in [5.67, 6.93]$, R_1, R_2, R_4, $R_6 \in [0.09, 0.11]$, R_3, $R_5 \in [1.8, 2.2]$. Applying nodal analysis we obtain following system of equations:

$$\begin{bmatrix} [1.98,\ 2.42] & [-2.2,\ -1.8] & [0,\ 0] \\ [-2.2,\ -1.8] & [3.69,\ 4.51] & [-2.2,\ -1.8] \\ [0,\ 0] & [-2.2,\ -1.8] & [1.89,\ 2.31] \end{bmatrix} \begin{bmatrix} V_1 \\ V_2 \\ V_3 \end{bmatrix} = \begin{bmatrix} [0.51,\ 0.76] \\ [0,\ 0] \\ [0,\ 0] \end{bmatrix}$$

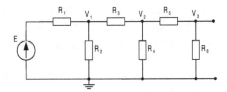

Figure 1. R-ladder circuit

It's seen that the coefficients and right-hand sides of the system are not determined exactly, but are only known to lie within some real intervals.

Denote a system of linear interval equations by

$$\mathbf{A}\, x = \mathbf{b}. \tag{1}$$

The coefficients and right-hand sides of the system are not determined exactly, but are only known to lie within some real intervals. Such a system of linear interval equations represents a family of ordinary linear systems which can be obtained from it by fixing coefficients and right-hand-side values in the prescribed intervals. Each of these systems, under the assumption that each $A \in \mathbf{A}$ is nonsingular, has a unique solution, and all these solutions constitute a so-called solution set S. The solution set of Eq.(1) can be expressed as

$$S = \{x : Ax = b,\ A \in A,\ b \in b\}. \tag{2}$$

It forms some multidimensional operating region of a circuit.

It the next points, we introduce some notations and ideas concerned with a structures of operating regions, drawbacks of some circuit descriptions, and method of rectangular (interval) evaluating of the operating regions.

As the examples, we give the calculation of variations of the frequency response and nodal voltages of the RC circuits with interval data. We also test our approach when the widths of matrix elements in Eq.(1) are large.

2.1. Notations and preliminaries

All vectors in this paper will have n components and all matrices will be of size n×n. The sets of real vectors, real matrices, interval vectors, and interval matrices are represented by the lower case, upper case, lower case bold, and upper case bold letters respectively. A real, scalar interval x is given by $[\underline{x}, \bar{x}]$, where the endpoints of an interval are $\underline{x} \leq \bar{x}$. We denote the center of an interval x by $m(\mathbf{x}) = (\underline{x} + \bar{x})/2$, and width of an interval by $w(\mathbf{x}) = \bar{x} - \underline{x}$. For interval vectors and matrices these concepts are defined via the elements. If $\mathbf{A} = (\mathbf{a}_{ij})$ is an interval matrix, then $m(\mathbf{A}) = (m(\mathbf{a}_{ij}))$, see for example [2]. We say that a real vector x, is contained in an interval vector $\mathbf{x}$, and we write $x \in \mathbf{x}$, if $\underline{x}_i \leq x_i \leq \bar{x}_i$ for all i = 1,...,n. A real matrix A is contained in an interval matrix $\mathbf{A}$, and we write $A \in \mathbf{A}$, if $\underline{a}_{ij} \leq a_{ij} \leq \bar{a}_{ij}$ for all i, j = 1,...,n. An interval matrix $\mathbf{A}$ is called regular if $\det A \neq 0$ for each $A \in \mathbf{A}$. Let $|A| = (|a_{ij}|)$ denote matrix with absolute values of elements and $A \geq 0$ (and similar relations) are meant componentwise (i.e. $a_{ij} \geq 0$). An interval matrix $\mathbf{A}$ is called inverse stable if $|A^{-1}| > 0$ for each $A \in \mathbf{A}$, i.e. if each inverse matrix element is nonzero [12]. Subsequently we introduce some concepts of matrix theory [11]. First, we use the (componentwise) natural partial ordering on sets of real rectors and matrices:

$$A \geq B \Leftrightarrow a_{ij} \geq b_{ij} \ \text{ for } i, \ j = 1, \ ..., \ n$$

For inverse stable matrix A the signature matrix Z is defined by

$$z_{ij} = \begin{cases} 1 & \text{if} \quad a_{ji} > 0 \\ -1 & \text{if} \quad a_{ji} < 0 \end{cases} \quad i, \ j = 1, \ ..., \ n$$

(notice the transposition of indices). For two n×n matrices A = (a_{ij}) and B = (b_{ij}) their componentwise product is defined as $A*B = (a_{ij}b_{ij})$. Further, diag A denote the diagonal vector of A, i.e. diag $A = (a_{11},...,a_{nn})^T$ and $\rho(A)$ denotes spectral radius of A.

2.2. Structure of an operating region

In the first section we introduce a linear interval $\mathbf{Ax} = \mathbf{b}$ with an interval matrix of coefficients $\mathbf{A} = [A-\Delta, A+\Delta]$ and a right-hand side interval vector $\mathbf{b} = [b-\delta, b+\delta]$ describing the linear circuit. Here A = m($\mathbf{A}$), Δ = w($\mathbf{A}$)/2 and b = m($\mathbf{b}$), δ = w($\mathbf{b}$)/2. The set S of all possible operating points (solutions of Eq.(1)) of linear circuit may have a very complicated structure. This set is not generally an interval vector.

A number of authors have described ways to compute an x satisfying (2). See [12] for reference. In their pioneer work Oettli and Prager [8] showed that the solution set of a linear interval equation with regular matrix A is described following

$$x \in S \Leftrightarrow |Ax - b| \le \Delta|x| + \delta \tag{3}$$

It is known [12] that S can be represented as a union of at most 2^n convex polyhedra. The intersection of S with each orthant of R^n is convex. In general, S itself is not convex.

EXAMPLE 1. To illustrate that the solution set S given by (3) is not simple, let consider a linear system with input-output relationship written as $y(j\omega) = K(j\omega,p) \, x(j\omega)$.

Frequency response $K(j\omega,p)$ vary with vector p of some parameters ranging in known intervals. It can be shown [9] that for a fixed frequency changes of $K(j\omega,p)$ are described by the system of two interval equations

$$\begin{bmatrix} [a,\ b] & -[c,\ d] \\ [c,\ d] & [a,\ b] \end{bmatrix} \begin{bmatrix} y_1 \\ y_2 \end{bmatrix} = \begin{bmatrix} [1,\ 1] \\ 0 \end{bmatrix}$$

where the intervals [a,b] and [c,d] represent the ranges of values of $\text{Re}\{K(j\omega,\ p)^{-1}\}$ and $\text{Im}\{K(j\omega,\ p)^{-1}\}$, respectively, and $K(j\omega,p) = y_1 + jy_2$. In Fig.2 is presented the region of frequency response changes caused by variations of system parameters done by intervals [a,b] = [1,2] and [c,d] = [-0.5,0.5]. Coordinates of the points fixing the shape of the region are following: A(4/9, 2/9), B(4/3, 2/3), C(1, 0), D(4/3, -2/3), E(4/9, -2/9), and F(1/2, 0).

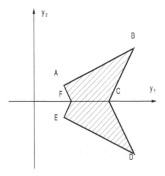

Figure 2. The region of changes of frequency response for a fixed frequency

Because S is generally so complicated in shape, it is usually impractical to try to use it. Instead, it is common practice to seek the interval vector x containing S which has the narrowest possible interval components. We say we "solve" the problem when we find x.

2.3. Rectangular evaluation of an operating region

From the standpoint of view of interval computations the formulation of circuit equations based of mesh analysis or modal analysis are not necessarily suitable for the networks with interval parameters. They result in a system of linear interval equations whose coefficients are not independent. We have the wider range of the coefficients since the elements of mesh-impedance matrix or mode-admittance matrix are given by the linear combinations of the interval parameters of resistors, capacitors, etc. Hence the width of the elements of the matrix coefficients and right-hand side vector of Eq.(1) becomes resultantly larger. This makes the evaluation of interval solution worse. Secondly owing to the interval dependency the linear combination of interval numbers gives us possibility to have the meaningless combination of the parameters. In order to avoid the drawbacks mentioned above the formulation by hybrid equation is well suited for solving linear interval systems [4], [10].

The problem treated in this section is how to compute the interval hull of S, i.e. the interval vector x with components

$$\begin{aligned} \underline{x}_i &= \min\{x_i, x \in S\} \\ \overline{x}_i &= \max\{x_i, x \in S\} \end{aligned} \quad i = 1,...,n \tag{4}$$

describing the exact range of components of the solution if coefficients and right-hand sides of Eq.(1) are allowed to vary in the given intervals.

Method of computing the vector $\mathbf{x} = [\underline{x}, \ \overline{x}]$ described by (4) is based on the results of Rohn [5], [6] slightly modified. Namely, the vectors of left endpoints and right endpoints of the interval (rectangular) evaluation of the interval hull are computed iteratively by

$$\begin{aligned} \underline{X}^{k+1} &= A^{-1}\left(B - Z*\left(\Delta\left|\underline{X}^k\right| + D\right)\right), \quad \underline{x} = diag\,\underline{X} \\ \overline{X}^{k+1} &= A^{-1}\left(B + Z*\left(\Delta\left|\underline{X}^k\right| + D\right)\right), \quad \overline{x} = diag\,\overline{X} \\ &\qquad k = 0,1,... \end{aligned} \tag{5}$$

Where $B = be^T$, $D = \delta e^T$, $e = (1,1,...,1)^T \in R^n$. Z is a signature matrix of A^{-1}. Recommended initial guess for the sequence (5) is

$$\underline{X}^o = \overline{X}^o = diag\,A^{-1}B. \tag{6}$$

It is assumed that coefficient matrix $\mathbf{A}$ is regular and inverse stable. It is assured if following conditions are satisfied

$$\rho\left(\left|A^{-1}\right|\Delta\right)<1 \tag{7}$$

and

$$\Delta\left(I-\left|A^{-1}\right|\Delta\right)^{-1}\left|A^{-1}\right|<I \tag{8}$$

(I is the unit matrix).

Conditions (7) and (8) under which the method works are satisfied if all coefficients of A^{-1} are nonzero and Δ is sufficiently small. Under this conditions the sequences $\{\underline{x}^k\}$, $\{\bar{x}^k\}$ converge from starting point (6). Note that each column of $\underline{X}^o$, $\overline{X}^o$ is equal to an approximate solution to Ax=b.

2.4. Numerical examples

EXAMPLE 2. Consider the simple RC voltage divider circuit of Fig.3, where the resistor and the capacitor values are allowed to vary from 4.4Ω to 4.6Ω and from 520 µF to 580 µF respectively.

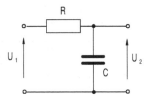

R

U_1 U_2

C

Figure 3. RC voltage divider

We want to calculate amplitude variations of the frequency response due to the changes of resistance and capacitance. For a fixed frequency, amplitude of the frequency response $K(\omega)=|U_2(j\omega)/U_1(j\omega)|$ is obtained as a function of variations of R and C by solving following two linear interval equations

$$\begin{bmatrix} 1 & -\omega[\underline{RC},\ \overline{RC}] \\ \omega[\underline{RC},\ \overline{RC}] & 1 \end{bmatrix}\begin{bmatrix} U_2^{(R)} \\ U_2^{(I)} \end{bmatrix}=\begin{bmatrix} 1 \\ 0 \end{bmatrix}$$

where $R=[\underline{R},\ \overline{R}]$, $C=[\underline{C},\ \overline{C}]$ and $U_1=1$, $U_2=U_2^{(R)}+jU_2^{(I)}$.

In Fig.4, we have plotted K(ω) as a function of ω. Hatched region show the result by Oettli-Prager condition. Outer bounds were computed with use of the sequences (5). Both results are in a fairly good agreement.

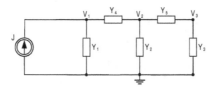

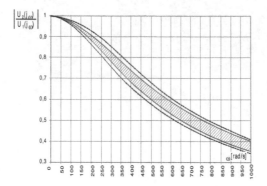

Figure 4. Variations of the frequency response of a voltage divider circuit

EXAMPLE 3. Consider the linear RC circuit illustrated in Fig.5 [9]. The interval parameter with the center value m and the width 2r is denoted as (m, r). The numerical data are following: $Y_k = G_k + jB_k$, $G_k = (1, \varepsilon)$, $B_k = (0.2, 0.2\varepsilon)$, J=10

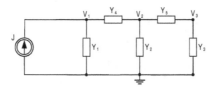

Figure 5. The circuit with complex admittances done as interval numbers.

In spite of remarks in previous section on drawbacks of nodal analysis we applied here nodal equations. It allowed us to fulfill assumptions of Rohn's algorithm although the elements of node-admittance matrix are dependent. In order to carry the sequences (5) working with Eq.(1) represented by real interval numbers we replaced the interval complex nodal equations

$$Y\ V = J$$

by the real interval equation

$$A\ x = b$$

where

$$A = \begin{bmatrix} G & -B \\ B & G \end{bmatrix}, \quad x = \begin{bmatrix} V_R \\ V_I \end{bmatrix}, \quad b = \begin{bmatrix} J_R \\ J_I \end{bmatrix}$$

$$J_R = Re(J), \quad J_I = Im(J)$$

$$G = Re(Y), \quad B = Im(Y), \quad V_R = Re(V), \quad V_I = Im(V)$$

The results are tabulated below

nodal voltage	$\varepsilon = 0.05$	
	Re	Jm
V_1	[5.7635, 6.2724]	[-1.2652, -11447]
V_2	[2.3055, 2.5090]	[-0.5061, -0,4579]
V_3	[1.1527, 1.2545]	[-0.2530, -0.2289]
nodal voltage	$\varepsilon = 0.1$	
	Re	Jm
V_1	[5.5336, 6.5529]	[-1.3355, -10927]
V_2	[2.2135, 2.6212]	[-0.5342, -0.4371]
V_3	[1.1067, 1.3106]	[-0.2671, -0.2185]

Table 1. Ranges for nodal voltages

Results are essentially the same as in [10] although the widths of nodal voltages are greater. It confirms the influence of coefficient dependence on evaluation of interval solutions.

EXAMPLE 4. To test properties of presented approach let consider the active two-port shown in Fig.6.

Current-controlled description of the two-port is following

$$\begin{bmatrix} u_1 \\ u_2 \end{bmatrix} = \begin{bmatrix} R_1 + R_3 - gR_1R_3 & R_3 + \alpha - gR_1R_3 \\ R_3 - \beta + gR_2R_3 & R_1 + R_3 + gR_1R_3 \end{bmatrix} \begin{bmatrix} i_1 \\ i_2 \end{bmatrix}$$

For numerical data

$$R_1 = R_2 = R_3 = 1, \quad \alpha = [-1,1], \quad \beta = 2, \quad g = [0,1]$$

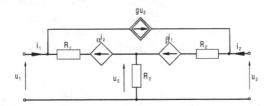

Figure 6. The active two-port with interval parameters.

interval representation is following

$$\begin{bmatrix} u_1 \\ u_2 \end{bmatrix} = \begin{bmatrix} [1, \ 2] & [0, \ 1] \\ [-1, \ 0] & [2, \ 3] \end{bmatrix} \begin{bmatrix} i_1 \\ i_2 \end{bmatrix}$$

If excitations are determined as $U_1 \in [2,3]$, $U_2 \in [0,1]$ the solutions set S has form shown in Fig.7.

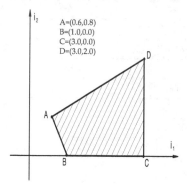

A=(0.6,0.8)
B=(1.0,0.0)
C=(3.0,0.0)
D=(3.0,2.0)

Figure 7. The solutions set of i_1 and i_2

Computing bounds on operating region for i_1 and i_2 with recommendation given by (6) and slightly different starting intervals following results were obtained:

$$i_1 = [\underline{i_1}, \ \overline{i_1}], \quad \underline{i_1} \in [-1.5, \ 0.3], \quad \overline{i_1} \in [3.9, \ 5.2]$$

$$i_2 = [\underline{i_2}, \ \overline{i_2}], \quad \underline{i_2} \in [-2.0, \ -1.2], \quad \overline{i_2} \in [2.5, \ 4.0]$$

It's seen that this approach give poor results if widths of matrix coefficients are large.

3. On finding operating points of nonlinear, inertialess circuits

A continuation method is a method of dc circuit analysis that under proper conditions, is guaranteed to find a circuit's operating point [26], [27]. It can be also used in calculating a circuit's periodic steady-state response [24], [25]. Finding the dc operating point is essential for circuit simulation: both steady-state and transient analyses require a priori knowledge of circuit's dc operating point. The idea behind a continuation method is to embed in the circuit's equations

$$F(x) = 0 \tag{9}$$

(nodal or hybrid) an additional parameter λ such that the circuit that corresponds to $\lambda = 0$ is trivial or easy to find: the circuit that corresponds to $\lambda = 1$ is the original circuit whose solution is desired. Hence, a mapping (homotopy) H: $R^{n+1} \rightarrow R^n$ is constructed, where H(x, 0) = 0 gives the solution to the initial (start) circuit and H(x, 1) = 0 gives the solution to the circuit being simulated. To find this solution, one generate a sequence of points $\{(x^k, \lambda^k)\}_{k=0}^N$, where x^k is on or near the path and x^0 is a known solution of H(x,0) = 0. The path in R^{n+1} begins at $\lambda = 0$ and ends at $\lambda = 1$. An important problem in following homotopy path is a control of the step from a point x^k to x^{k+1}. The design philosophies of a step size control are numerous and varied. See, e.g. [28] - [31].

Most of the methods on the following continuation paths are based on approximate step control. Approximate step control methods are successful and fast when the path is smooth and isolated, but problems arise when there are many paths near some points. In that case, algorithms based on approximate step control methods may jump from one path to another. Also, if rapid changes in curvature occur along the path, the method based on approximate step control sometimes even erroneously reverses orientation. However, appropriate use of interval analysis gives us guarantee that the predictor algorithm will not jump from one path to another, or, indeed, jump over different legs of the same path. We present an interval step control for tracing continuation paths which assures that the predictor-corrector iterations will not jump across paths, and each predictor step is as large as possible, subject to verification that the path is unique with the given interval extension.

3.1. Step size control

To trace a path from a known solution (x^0, λ^0), we first predict the solution for $\lambda = \lambda^0 + \Delta\lambda$ and then correct the prediction using Newton method with λ fixed. Specifically, for small Δx and $\Delta\lambda$, the Taylor expansion for H gives

$$H(x^0 + \Delta x, \lambda^0 + \Delta\lambda) \approx H(x^0, \lambda^0) + J_x(x^0, \lambda^0)\Delta x + J_\lambda(x^0, \lambda^0)\Delta\lambda \qquad (10)$$

where $J_x(x^0, \lambda^0)$ and $J_\lambda(x^0, \lambda^0)$ are the Jacobian matrices of H with respect to x and λ. For the prediction step, we have $H(x^0, \lambda^0)=0$, so setting $H(x^0 + \Delta x, \lambda^0 + \Delta\lambda)=0$ gives

$$\Delta x = -J_x(x^0, \lambda^0)^{-1} J_\lambda(x^0, \lambda^0)\Delta\lambda. \qquad (11)$$

Since this is only approximate, we correct the solution at the new value of $\lambda = \lambda^0 + \Delta\lambda$ by Newton-Raphson algorithm starting with initial guess $x_0 = x^0 + \Delta x$. A common numerical practice is to stop the Newton iteration whenever the distance between two iterates is less than a given tolerance, i.e., when

$$\left\| x_{k+1} - x_k \right\| < \varepsilon \qquad (12)$$

However, just the fact that (12) is satisfied does not guarantee the existence of a solution. We try to overcome this difficulty by first performing the three Newton steps and using them to compute an interval vector that is very likely to contain a solution of the correction step. We adopt here results presented in [22]. Consider Newton's method

$$x_{k+1} = x_k - J_x(x_k, \lambda)^{-1} H(x_k, \lambda), \quad x_0 = x^0 + \Delta x, \quad k = 0, 1., \tag{13}$$

Let

$$\rho_k = \left\| x_{k+1} - x_k \right\|_\infty, \tag{14}$$

for some fixed k and

$$\mathbf{x} = \{ x \in R^n \left\| x_1 - x \right\|_\infty < \rho_0 \} \tag{15}$$

Combining the properties of the Krawczyk operator and a corollary of the Newton-Kantorovich theorem it can be shown, that for three successive iterates the following relation holds

$$\frac{8\rho_1^3}{\left(\left\| x_2 \right\|_\infty \cdot \rho_0^2 \left(\left\| x_2 \right\|_\infty \cdot \rho_0^2 \right) \right)} < eps. \tag{16}$$

In Fig. 8 there is a sketch of the idea of constructing the predictor - corrector step for two-dimensional problems, i.e. x = (y, z).

We can compute interval (15) and using the Krawczyk operator K(x) we test whenever this interval contains a solution. It is true if the Krawczyk operator satisfy an inclusion

$$K(\mathbf{x}) = x_2 - C \cdot H(x_2, \lambda) + (I - C \cdot J_x(\mathbf{x}, \lambda))(\mathbf{x} - x_2) \subseteq \mathbf{x} \tag{17}$$

then x contain solution of the correction step x^*. Here $J_x(x, \lambda)$ denotes interval arithmetic evaluation of $J_x(x, \lambda)$, $C = [J_x(x_1, \lambda)]^{-1}$, and I is identity matrix [20]. If additional condition

$$\left\| I - C \cdot J_x(\mathbf{x}, \lambda) \right\| < 1 \tag{18}$$

is satisfied [21], then the solution x^* is unique in x and the operator (17) generates a sequence of intervals x_r, $r = 1, 2, \ldots, m$, that satisfy the relations

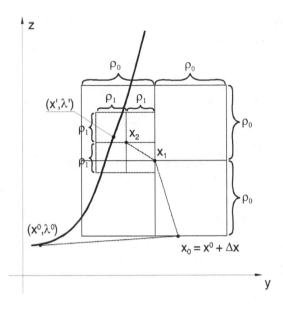

Figure 8. The predictor-corrector step for two dimensional problem

$$x^* \in \mathbf{x}_m \subseteq \mathbf{x}_{m-1} \cdots \qquad (19)$$

This nested sequence is stopped when the width of the interval x_m becomes smaller than

$\mu > 0$ and then $x^* = m(x_m)$. The correction step may be repeated several times before taking the next prediction step.

Step size control can be formalized in the following algorithm.

Algorithm

Unless otherwise indicated STEP {k+1} follows STEP {k}.

Given: Solution (x^0, λ^0) of a homotopy $H(x, \lambda) = 0$.

STEP {1} Assume $0 < \Delta\lambda < 1$;

STEP {2} Compute prediction step Δx according to (11);

STEP {3} Compute using Newton method (13) three successive iterates;

STEP {4} Verify the inequality (16) for some eps "/> 0;

STEP {5} Check the relation (17). If (17) is not satisfied take $\Delta\lambda = \Delta\lambda/2$ and go to {2};

STEP {6} Compute interval **x** according to (14) and (15);

STEP {7} Check the condition (18);

STEP {8} Generate a sequence (19) according to

$\mathbf{x}_1 = K(\mathbf{x})$, $\mathbf{x}_2 = K(\mathbf{x}_1)...\mathbf{x}_m = K(\mathbf{x}_{m-1})$

until $w(\mathbf{x}_m) < \mu$, $\mu > 0$,

where $w(\mathbf{x}_m) = \check{x}_m - x_m$ (component wise);

STEP {9} Compute

$x^* = m(\mathbf{x}_m) = (x_m + \check{x}_m) / 2$ (component wise);

STEP {10} Assign $x^0 \leftarrow x' = x^*$ and $\lambda^0 \leftarrow \lambda'$

and go to {1} (new prediction);

3.2. Box — Bisection searching

Introducing some interval operator (e.g. Krawczyk operator, Hansen-Sengupta operator or some of their modifications [20]) for system of nonlinear equations we can formulate generalized bisection algorithm applicable to find solutions of equation (9).

Let denote Krawczyk operator associated with Eq. (9) as

$$K(\mathbf{x}, y, F) = y - YF(y) + R(\mathbf{x})(\mathbf{x} - y) \tag{20}$$

Here $Y = [m(F'(\mathbf{x}))]^{-1}$ where $F'(\mathbf{x})$ is an interval arithmetic evaluation of $F'(x)$ (Jacobian matrix of Eq. (1)), $y \in \mathbf{x}$ (e.g. y = m(\mathbf{x})),

$$R(\mathbf{x}) = I - YF'(\mathbf{x}) \tag{21}$$

and I is identity matrix.

Let T denotes the list of subregions yet to be tested. P is the list of subregions of B which may contain a solution to (9) but which are too small for further analysis, i.e. $w(\mathbf{x}) \le \varepsilon$ (ε - accuracy of searching).

Unless otherwise indicated STEP {k+1} follows STEP {k}.

Given: n-dimensionsl box (interval vector) $B \subseteq D \subseteq R^n$.

Algorithm

STEP {1} Assign $\mathbf{x} \leftarrow B$; T $\leftarrow \emptyset$; P $\leftarrow \emptyset$;

STEP {2} Compute F(\mathbf{x});

STEP {3} If $0 \notin F(\mathbf{x})$ go to STEP {11};

STEP {4} Compute Y, if Y is singular go to STEP {9};

STEP {5} Assuming y = m(**x**) compute K(**x**, y, F) and R(**x**);

STEP {6} If K(**x**, y, F) ∩ **x** = ∅ go to STEP {11};

STEP {7} If ||K(**x**, y, F)-**x**||≤w(**x**)/2|| , then **x** contains a solution, continue, else go to STEP {9};

STEP {8} ||R(**x**)<1|| If , then there is a unique solution to (1) in **x**, if w(**x**) ≤ ε terminate search, go to STEP {11};

STEP {9} If w(**x**) ≤ ε add **x** to the list P and go to STEP {11} else bisect **x** on **x**' and' **x**'' according to some rules of bisection;

STEP {10} Assign **x** ← **x**', add **x**'' to the head of the list T, go to STEP {2};

STEP {11} If list T is empty go to STEP {12}, otherwise take an interval vector **x** $_p$ at the head of list T, assign **x** ← **x** $_p$, delete this vector from list T and go to STEP {2};

STEP {12} If list P is empty, terminate search with no solution in B, otherwise print list P and terminate search.

One can simply check that in the STEPS {5}, {7}, {8} for Algorithm with step size control and in the STEPS {2}, {4},{5} for box – bisection searching one needs to evaluate the ranges of nonlinear functions over some interval **x**. This problem can be solved by computing coefficients of Bernstein polynomials. For details see [34].

3.3. Numerical examples

As an illustration of the implementation of the approach outlined above, consider the following two examples.

EXAMPLE 5. Our first example concerns a simple circuit of Fig. 9 which consists of two tunnel diodes and two constant current sources and a linear resistor. The tunnel diode characteristics are defined following

$$h_1(u_1)=h_2(u_2)=h\,(u)=5.0u-1.7u^2+0.15u^3$$

The network equations are obtained from the hybrid analysis as

$$F(x)=F(u_1,\,u_2)=\begin{bmatrix}h\,(u_1)\\h\,(u_2)\end{bmatrix}+\frac{1}{R}\begin{bmatrix}1&1\\1&1\end{bmatrix}\begin{bmatrix}u_1\\u_2\end{bmatrix}-\begin{bmatrix}j_1\\j_2\end{bmatrix}=\begin{bmatrix}0\\0\end{bmatrix}$$

We introduce homotopy of the form

$$H(x,\,\lambda)=\gamma(1-\lambda)Q(x)+\lambda F(x)$$

where $\lambda \in [0,1]$ and γ is a random complex number, different from zero.

Start equations are following

$$Q(x)=Q(u_1,\,u_2)=\begin{cases}u_1^3-1\\u_2^3-1\end{cases}$$

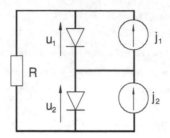

Figure 9. The circuit of two tunnel diodes

It's seen that the solutions of equation

$$H(x,\ 0)=Q(x)=0$$

are distributed on unit circles and they form nine starting points $(x^0, 0)$. For solving so prepared homotopy equations we splited them into real and imaginary parts.

Presented approach leads to the five real solutions

$$Q_1=(1.805435,\ 7.325956),$$
$$Q_2=(3.609712,\ 6.937570),$$
$$Q_3=(6.103275,\ 6.103273),$$
$$Q_4=(6.937570,\ 3.609712),$$
$$Q_5=(7.325956,\ 1.805434).$$

Because above equation is equivalent polynomial of 9th degree there are also four complex solutions:

$$Q_6=(1.493997 + j2.483038,\ 1.493997 + j9.932153),$$
$$Q_7=(1.493997 - j2.483038,\ 1.493997 - j9.932153),$$
$$Q_8=(2.615029 + j2.812082,\ 2.615029 + j1.687249),$$
$$Q_9=(2.615029 - j2.812082,\ 2.615029 - j1.687249).$$

EXAMPLE 6. Let consider circuit hybrid equation [17].

$$F(x)=\begin{bmatrix} f_1(x_1) \\ f_2(x_1,\ x_2) \\ f_3(x_2,\ x_3) \end{bmatrix} - \begin{bmatrix} 1 & 1 & 1 \\ 0 & -1 & 0 \\ 0 & 1 & 1 \end{bmatrix}\begin{bmatrix} x_1 \\ x_2 \\ x_3 \end{bmatrix} - \begin{bmatrix} -5 \\ 5 \\ -5 \end{bmatrix} = \begin{bmatrix} 0 \\ 0 \\ 0 \end{bmatrix}$$

Nonlinear functions (characteristics of nonlinear elements) are following

$$f_1(x_1)=x_1^3$$
$$f_2(x_1,x_2)=0.25x_1x_2$$
$$f_3(x_2,x)=0.048(x+x).$$

For solving this polynomial system we have applied the same homotopy as in Example 5.

Using concept of m-homogenity [18],[19] we can introduce solutions that corresponds to $\lambda=0$ and the start system $H(x, 0) = Q(x) = 0$ look like this

$$Q(x) = \begin{cases} -0.509x_1^3+1.235x_1^2-2.961x_1x_2+0.902x_1^2x_2 \\ +0.078x_1x_3+0.196x_1+2.294x_2-0.235x_3+1=0 \\ -0.228x_1x_2-0.024x_1x_3+0.492x_1+0.443x_2 \\ +0.069x_3+1=0 \\ -0.641x_2^2-0.143x_3^2+0.642x_2x_3+1.603x_2 \\ -0.781x_3+1=0 \end{cases}$$

The start system respects the m-homogeneous structure [18] of the hybrid equation we want to solve and has eight start solutions. Namely

$$x_1^0=(3.0, 2.0, 1.0), \quad x_2^0=(1.5, 2.0, 1.0),$$
$$x_3^0=(2.0, 3.02.5,), \quad x_4^0=(2.5, 1.5, 1.0),$$
$$x_5^0=(1.0, 2.0, 2.5), \quad x_6^0=(2.0, 1.0, 3.0),$$
$$x_7^0=(1.5, 1.0, 2.0), \quad x_8^0=(3.0, 2.0, 3.0).$$

We applied the step size control previously described with tolerance eps = 0.01. We needed to trace only eight solution paths. Two pairs of them reached the same solutions, and so we obtained following six solutions of the hybrid equations:

$$x_A=(1.7155, 3.4992, 4.8340),$$
$$x_B=(2.1273, 3.2641, 9.2359),$$
$$x_C=(-0.8578 + j1.0988, 5.6714 - j1.9832, 2.6619 + j1.9832),$$
$$x_D=(-0.8578 - j1.0988, 5.6714 + j1.9832, 2.6619 - j1.9832),$$
$$x_E=(-1.9832 + j1.5473, 5.3309 - j2.8091, 7.1691 + j2.2091),$$
$$x_F=(-1.9832 - j1.5473, 5.3309 + j2.8091, 7.1691 - j2.2091).$$

For circuit analysis only real solutions x_A and x_B are of interest.

EXAMPLE 7. In the last example an electric circuit depicted in Fig. 10 assembling some linear resistors, diodes and operational amplifier is considered (cf. [29]).

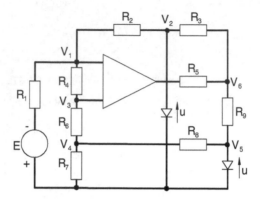

Figure 10. The circuit with operational amplifier and diodes

The voltage induced in the operational amplifier is modeled by:

$$u_A(x)=7.65\,\mathrm{arctan}(1962x),$$

the current through the diodes satisfies:

$$i_D(u)=5.6\cdot 10^{-8}\!\left(e^{25u}-1\right);$$

resistances in ohms are following: $R_1 = 51$, $R_2 = 39$, $R_3 = 10$, $R_4 = 10^4$, $R_5 = 0.201$, $R_6 = 25.5$, $R_7 = 0.62$, $R_8 = 1$, $R_9 = 13$. Applying Kirchhoff's laws to the circuit leads to the system of nonlinear equations:

$$f_1=(v_1-v_3)/R_4+(v_1-v_2)/R_2+(v_1+E)/R_1=0$$
$$f_2=(v_2-v_6)/R_3+i_D(v_2)+(v_2-v_1)/R_2=0$$
$$f_3=(v_3-v_1)/R_4+(v_3-v_4)/R_6=0$$
$$f_4=(v_4-v_3)/R_6+v_4/R_7+(v_4-v_5)/R_8=0$$
$$f_5=(v_5-v_4)/R_8+(v_5-v_6)/R_9+i_D(v_5)=0$$
$$f_6=(v_6-v_2)/R_3-[u_A(v_3-v_1)-v_6]/R_5$$
$$+(v_6-v5)/R_9=0$$

Functions describing nonlinearities of diodes and operational amplifier are smooth functions so we have needed their expansion to compute range values. We have assumed Taylor expansion with degree = 4 and for Bernstein coefficients degree = 10 [34]. Applying box–bisection algorithm we have obtained solutions for two different ranges of the input voltage E. Results are tabulated as

E	[0.320, 0.322]	[0.599, 0.601]
v_1	[0.2325, 0.2354]	[0.0439, 0.0493]
v_2	[0.6578, 0.6629]	[0.5423, 0.5471]
v_3	[0.2303, 0.2375]	[0.0464, 0.0492]
v_4	[0.2299, 0.2375]	[0.0446, 0.0496]
v_5	[0.6155, 0.6208]	[0.1232, 0.1293]
v_6	[9.5941, 9.6088]	[1.1597, 1.1662]

Table 2. Solutions presented here are in good agreement with the results of [29].

4. Concluding remarks

In first section, we applied an algorithm for bounding the solution set of a system of linear interval equation which works when the interval elements of **A** and the interval components of **b** are relatively narrow. In practice, however, we are often in situation like in Example 4 when the widths of these intervals are rather wide. This case creates difficulties. Presented here version of Rohn's algorithm doesn't work well. Gaussian elimination using interval arithmetic tends to give poor results. Interval Gaussian elimination can fail because of division by an interval containing zero. This can occur even when the interval coefficient matrix is regular i.e. does not contain a singular real matrix [2]. Maybe methods based on preconditioning of the linear interval equations [7] offer new possibilities. However, recently it was reported that the problem of computing the exact bounds of solution set for a system of linear interval equations is NP-hard (computationally intractable) [13]. Shortly speaking, NP-hardness of a problem P means that if we are able to solve this problem in reasonable time, then we would be able to solve all problems from a very large class of complicated problems (called class NP) in reasonable time, and this possibility is widely believed to be impossible. Reasonable time means a time that does not exceed some polynomial of the length of the input. For exact definitions see, e.g., [15]. On the other hand it was also proved [14] that if the interval components of **A** and **b** are „thin" enough, then there exists a polynomial-time algorithm that computes the exact bounds for S in „almost all" cases („almost all" in some reasonable sense). It means that solving linear interval equations isn't a hopeless task.

Second sections presents an approach to compute approximations to the solution path of a systems of nonlinear equations. The suggested method is based on predictor-corrector principle. We have combined common Newton-Raphson correction step with the properties of the interval Krawczyk operator. An interval step control for tracing continuation path assures that the predictor-corrector iterations were not jumping across paths. A robust and efficient path tracker was obtained by halving the homotopy parameter step whenever uniqueness of correction solution is not assured within a given tolerance of the continuation path and doubling it when the prediction step has been sufficiently accurate several consecutive times. Presented here path tracker was more effective than the tracing homotopy path via integration so-called basic differential equations [23]. Krawczyk operator was also useful in generalized box-bisection for searching all solutions of the circuit equations.

Author details

Zygmunt Garczarczyk*

Silesian University of Technology, Gliwice, Poland

References

[1]   Moore, R. E. Methods and Applications of Interval Analysis, SIAM Studies 2. Philadelphia: SIAM; (1979).

[2]   Alefeld, G, & Herzberger, J. Introduction to Interval Computations. New York: Academic Press; (1983).

[3]   Spence, R, & Soin, R. S. Tolerance Design of Electronic Circuits. London: Addison-Wesley; (1988).

[4]   Kolev, L. V. Interval Methods for Circuit Analysis. Singapore: World Scientific; (1993).

[5]   Rohn, J. Systems of linear interval equations. Linear Algebra Appl. (1989). , 126-39.

[6]   Rohn, J. A two-sequence method for linear interval equations. Computing (1989).

[7]   Hansen, E. R. Bounding the solution of interval linear equations. SIAM J. Numer Anal. (1992). , 29(5), 1493-1503.

[8]   Oettli, W, & Prager, W. Compatibility of approximate solution of linear equations with given error bounds for coefficients and right-hand sides. Numer. Math. (1964). , 6(1), 405-409.

[9]   Garczarczyk, Z. Computing componentwise bounds on operating regions of linear circuits with interval data. In Proceedings of the 13th European Conference on Circuit Theory and design, ECCTD'97, 30 August- 3 September (1997). Budapest.

[10] Okumura, K, & Higashino, S. A method for solving complex linear equation of AC network by interval computation. In Proceedings of the 1994 International Symposium on Circuits and Systems, ISCAS'94, 30 May- 2 June (1994). London.

[11] Varga, R. S. Matrix Iterative Analysis, Englewood Cliffs,NJ: Prentice-Hall, (1962).

[12] Neumaier, A. Interval Methods for Systems of Equations. Cambridge: Cambridge University Press, (1990).

[13] Klatte, R, et al. PASCAL-XSC. Berlin: Springer Verlag, (1992).

[14] Rohn, J, & Kreinovich, V. Computing exact componentwise bounds on solution of linear systems with interval data is NP-hard. SIAM J. Matrix Anal. Appl. (1992). , 16(2), 415-420.

[15] Lakeyev, A. V, & Kreinovich, V. If input intervals are small then interval computations are almost always easy. Reliable Computing 1995, Supplement (1995). , 134-139.

[16] Garey, M. E, & Johnson, D. S. Computers and Intractability: A Guide to the Theory of NP-Completeness. San Francisco: Freeman, (1979).

[17] Chua, L. O, & Lin, P. M. Computer-Aided Analysis of Electronic Circuits. Englewood Cliffs, NJ: Prentice-Hall, Inc., (1975).

[18] Morgan, A, & Sommese, A. A homotopy for solving general polynomial systems that respects m-homogeneous structures. Appl.Math.Comput. (1987). , 24(2), 101-113.

[19] Garczarczyk, Z. Circuit design problems via polynomial equations solving, In Proceedigs of the 12th European Conference on Circuit Theory and Design, ECCTD'August (1995). Istanbul, Turkey., 95, 27-31.

[20] Moore, R. E, Kearfott, R. B, & Cloud, M. J. Introduction to Interval Analysis. Philadelphia: SIAM, (2009).

[21] Moore, R. E. A test for existence of solutions to nonlinear systems. SIAM J. Numer. Anal. (1977). , 14(4), 611-615.

[22] Alefeld G, Gienger A, & Potra F. Efficient numerical validation of solutions of nonlinear systems. SIAM J. Numer. Anal. (1994).

[23] Garcia, C. B, & Zangwill, W. I. Pathways to Solutions, Fixed Points and Equilibria. Englewood Clifs NJ: Prentice-Hall, Inc., (1981).

[24] Lu, X, Li, Y, & Su, Y. Finding periodic solutions of ordinary differential equations via homotopy method. Appl. Math. Comput. (1996). , 78(1), 1-17.

[25] Trajkovic, L, Fung, E, & Sanders, S. HomSPICE: simulator with homotopy algorithms for finding dc and steady-state solutions of nonlinear circuits. In Proceedings of the 1998 International Symposium on Circuits and Systems, ISCAS'98, 31 May- 3 June (1998). Monterey, CA.

[26] Trajkovic, L, & Willson, A. N. Theory of dc operating points of transistor networks, Intern. J. Electronics and Comm. (1992). , 46(4), 228-240.

[27] Mathis, W, et al. Parameter embedding methods for finding dc operating points of transistor circuits. In Proceedings of the International Specialist Workshop on Nonlinear Dynamics of Electronics Systems, NDES'95, July (1995). Dublin, Ireland.

[28] Allgower, E. L, & Georg, K. Numerical Continuation Methods: An Introduction. New York: Springer Verlag; (1990).

[29] Seydel, R, & Hlavacek, V. Role of continuation in engineering analysis. Chemical Engineering Science (1987). , 42(6), 1281-1295.

[30] Forster, W. Some computational methods for systems of nonlinear equations and systems of polynomial equations. J. Global Optimization (1992). , 2(4), 317-356.

[31] Lundberg, B. N, & Poore, A. B. Variable order Adams-Bashforth predictors with an error-stepsize control for continuation methods. SIAM J. Sci. Stat. Comput. (1991). , 12(3), 695-723.

[32] Garczarczyk, Z. On step size control for tracing continuation paths. In Schwarz W. (ed.) Proceedings of the 7th International Specialist Workshop on Nonlinear Dynamics of Electronics Systems, NDES'July, Ronne, Island of Bornholm, Denmark. Dresden: Technical University Dresden;(1999). , 99, 15-17.

[33] Garczarczyk, Z. A method for evaluation the range values of a bivariate function. In Proceedings 1998 International Symposium on Nonlinear Theory and its Applications, NOLTA'September (1998). Crans-Montana, Switzerland., 98, 14-17.

[34] Garczarczyk, Z. Linear Analog Circuits Problems by Means of Interval Analysis Techniques. In Tlelo-Cuautle E. (ed.) Advances in Analog Circuits. Rijeka: InTech; (2011). , 2011, 147-164.

Chapter 4

Memetic Method for Passive Filters Design

Tomasz Golonek and Jantos Piotr

Additional information is available at the end of the chapter

1. Introduction

The design of analog passive filters with specialized (not typical) frequency responses is not a trivial problem. The presence of finite load impedances for filter sections and limited quality factors of coils are just two of many concerns which a design engineer has to take into account. Additionally, classical techniques of filters synthesis require assuming of the approximation type (e.g. Butterworth, Chebyshev) before calculating the filter transfer function's poles and zeroes. This choice is frequently a challenge itself.

One of the methods allowing for elimination of the mentioned problems is the use of evolutionary computations (EC). Evolutionary techniques are a well known and frequently used tool of global optimization [1-3]. This kind of the optimization imitates natural processes of individuals' competition as candidates for reproduction. Better fitted individuals have higher survival probability and their genetic material is preferred. During the recombination process some parts of parents' genotypes are exchanged and offspring individuals are created. A new generation collected after the succession procedure conserves the features consisted in the previous genotypes. Besides, to assure a system resistance for a stagnation effect, mutation operations are applied to EC. The most popular sorts of EC approaches are: genetic algorithm (GA), genetic programming (GP), evolutionary strategies (ES), differential evolution (DE) and gene expression programming (GEP).

The main drawback of evolutionary approaches is an ineffective and insufficient local optimization. This property and significant computational efforts necessary for a huge generation processing predispose the EC applicability especially to the NP hard global searching problems [4-10]. This chapter describes the passive filters synthesis method by means of EC. The process of circuits' automated designing is a very complex issue. A wide area of solutions should be probed during the early stage of computations and its local parameters should be finally optimized. In contrary to the alternative systems [4-6], the method presented in this chapter is based on an application of a hybrid system - a synergy of genetic programming (GP) (used for the purpose of determining an optimal network of a passive filter circuit) and a deterministic local search by the means of Hooke and Jeeves method (HJM), which enables the system to find accurate values of the filter's elements. The proposed design system allows for obtaining the desired frequency response and, optionally, production yield optimization.

Section 2 explains the general algorithm of the proposed system, Sections from 3 to 5 present the descriptions of the important details of the algorithm. Next, in Section 6, the exemplary results of an automated circuit design are placed. Finally, in Section 7, some considerations of the method future development and final conclusions are presented.

2. Optimization process overview

The process is initialized with the desired filter specifications. Additional algorithms' parameters, e.g. population size, number of Monte Carlo analyses, and the like, are assumed. The use of GP and HJM is briefly presented in following paragraphs.

In the presented research both filters topology and circuit parameters values are being optimized. As far as GP has been proven an effective tool of circuits networks determination, it is not as efficient in adjusting resistors, coils, or capacitors values. The latter has been solved by the use of a deterministic, non-gradient local search algorithm - Hooke and Jeeves direct search method [11]. A synergy of evolutionary global optimization and local optimization algorithm is called a memetic algorithm [12-16]. In the presented research the proposition of the memetic genetic programming (MGP) introduction for the purpose of the analog filters design is described.

The block diagram of the optimization process has been presented in Fig. 1. After system running, the first, primary generation of the N_{mx} individuals is created randomly. It is very important to assure the possible wide range of dispersion for the starting solutions, so the diversity of this population is extremely desired. In the proposed system, the uniform probability of the primary individuals' randomization with the maximal allowed size limitation is applied. To evaluate the actual, random solutions, the fitness function is determined for each individual from population. The distances between the parameters of the evaluated phenotypes and the target specifications are checked during this process. Next, the GP system is executed. During reproduction, the mating pool is collected with the reproduction method that prefers more fitted individuals. An adequate strength of selection pressure has to be kept during this process, and it has crucial impact for a system convergence. During recombination process, pairs selected from an intermediate pool are crossed with assumed probability. Besides, offspring genotypes can be mutated and it assures adequate veracity of the population and allows for achieving the new regions of the searching space. Detailed information about the genetic operations and the fitness function of the GP part of the proposed memetic system are included in Section 4.

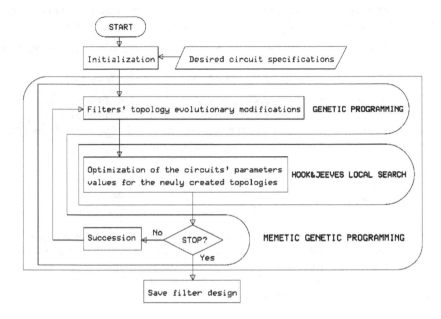

Figure 1. The process overview

This heuristic stage impacts on the coded circuits topology especially (i.e. the values of the filter resistances, inductances and capacitances are optimized not effectively), however the values of its elements are adjusted on the next stage during local deterministic optimization.

The newly created population of solutions represents a sort of circuits with non optimal values of its elements and now they are determined by means of HJM (pattern search) algorithm.

The differential of the fitness function is unknown. Hence, the local search algorithm could not have required its gradient computation. Among various optimization methods that meet this requirement, the pattern search method is characterized by its simplicity. The HJM method consists of two moves, i.e. exploratory and pattern, repeated sequentially until stop conditions are fulfilled.

Evolutionary algorithms are a trade-off between a global and local optimization. As far as they provide a good solution it never is but a sub-optimal one. The aim of the presented research was to achieve the best possible solution. Hence, it has been decided to use local search algorithms. The application of a local search algorithm after the evolutionary optimization is completed would improve the performance of the last chosen individual. Though, it would not affect the optimization process itself. Memetic solutions aim to improve the overall optimization process. The local search algorithm is applied within the evolutionary algorithm's main loop. It allows for faster convergence of the process. Applying a full local search process within each of the evolutionary iterations would influence the required computation time significantly and negatively. Therefore, only a few iterations of the HJM were allowed. For the purpose of improving the final result a full HJM cycle is applied after the evolutionary cycle is over.

In the presented approach the Hooke and Jeeves method had to be modified so the search is carried out in the assumed search space (E24 sequence of resistances, capacitances and inductances). The detailed description of the algorithm is presented in Section 5.

Next, after the topology and the values of elements determination, if the production yield optimization is required, the Monte Carlo analysis can be applied. The algorithm is terminated if all specification (and yield) requirements are met or if the last allowed generation G_{mx} was reached.

3. Circuit structure coding

The filter circuit structure is coded with the use of a binary tree structure. An exemplary tree is illustrated in Fig. 2. Its inner nodes contain functions, however terminals keep their arguments. The length from a root node to the most remote leaf is defined by the maximal depth D_{mx} and it reaches the value D_{mx}=4 for the illustrated case. Finally, the tree presented in Fig.2 can be decoded as a general expression given below:

$$y = f_1\left(f_2\left(t_1, f_3\left(t_2, t_3\right)\right), f_4\left(t_4, t_5\right)\right). \tag{1}$$

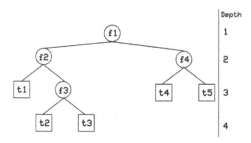

Figure 2. The example of tree structure

This kind of structure allows for defining the circuit topology and values of its elements in the flexible way with the assumed type of connection functions defined in **F** and the set of basic terminal blocks defined in **T**. Besides, it enables an easy way to the genetic operations implementation in the GP system. The set F contains unique symbols for connection types and for the described system it was defined as follows:

$$\mathbf{F} = \{—, >>, | \, |\}, \tag{2}$$

where contained symbols denote a serial, a cascade and a parallel configuration respectively.

Generally, passive filters circuits are constructed by the use of resistors, capacitors and magnetic coils (inductors), so the T-typed three-terminal (two-port) blocks which contain these elements (one in each branch) were chosen as basic ones:

$$\mathbf{T} = \left\{ Rv_e o_u Lv_e o_u Cv_e o_u, Lv_e o_u Cv_e o_u Rv_e o_u, Cv_e o_u Rv_e o_u Lv_e o_u \right\}, \tag{3}$$

where integer numbers v_e for $e=(1,..,E)$ and o_u for $u=(1,..,U)$ determine the value and its order for the element identified by the antecedent descriptor (i.e. R for a resistor, L for a inductor and C for a capacitor). In the proposed system, discrete values v_i are selected from a set $\mathbf{V}$ that contains elements from a widely known practical series E24 ($E=24$) prepared for the components manufactured with tolerance equal to $\delta_{tol}=5\%$:

$$\mathbf{V} = \left\{ \begin{array}{l} 10, 11, 12, 13, 15, 16, 18, 20, 22, 24, 27, 30, \\ 33, 36, 39, 43, 47, 51, 56, 62, 68, 75, 82, 91 \end{array} \right\}. \tag{4}$$

Order values are included in the set $\mathbf{O}$ and they should enable the desired range of coding, and consist of U integer ciphers ($0 \leq o_u \leq 9$) which determine exponents of decimal multipliers:

$$\mathbf{O} = \left\{ o_0, ..., o_{(U-1)} \right\}. \tag{5}$$

Finally, values of resistance R, inductance L and capacitance C are calculated from the equations (6) ÷ (8) respectively:

$$R = v_e \cdot 10^{o_u} [\Omega], \tag{6}$$

$$L = v_e \cdot 10^{o_u} [\mu H], \tag{7}$$

$$C = v_e \cdot 10^{o_u} [pF]. \tag{8}$$

The idea of the terminals' T (basic circuit blocks) coding is illustrated in Fig.3, where the three exemplary two-ports and adequate coding strings are presented. Impedances of elements from two-ports branches of these blocks are calculated from:

$$Z_R = R, \tag{9}$$

$$Z_L = 2\pi f L, \tag{10}$$

$$Z_C = \frac{1}{j2\pi f C}, \tag{11}$$

where f is a signal frequency and j denotes imaginary unit. Impedances Z_1, Z_2, Z_3 for terminal illustrated in Fig.3a can be determined according to equations (9), (10), (11) and its element sequence (i.e. $Z_1=Z_R$, $Z_2=Z_L$, $Z_3=Z_C$ for RLC-typed leaf, $Z_1=Z_L$, $Z_2=Z_C$, $Z_3=Z_R$ for LCR-typed leaf and $Z_1=Z_C$, $Z_2=Z_R$, $Z_3=Z_L$ for CRL-typed leaf).

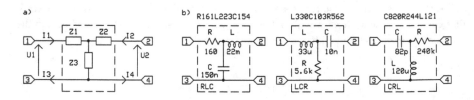

Figure 3. Terminal block structures: a) general, b) exemplary two-ports coded by the respective above strings

Finally, basing on elements from (2) and (3), the trees which code passive filters circuits can be constructed and some example is presented in Fig.4.

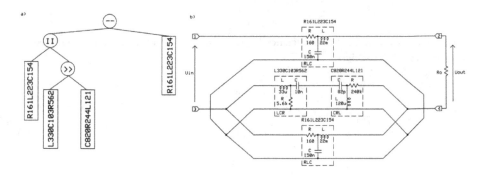

Figure 4. Exemplary structures of: a) genotype, b) phenotype

3.1. Analysis of the blocks connections

An important capability of the technique described is that it assures galvanic (direct) connection of the ground line between input and output ports and it is very desired for practically implemented filter circuits. Besides, as can be seen in the analysis below, each type of the basic blocks configurations allows for defining an equivalent T-typed circuit and it unifies the interpretation of genotypes of any shapes and sizes.

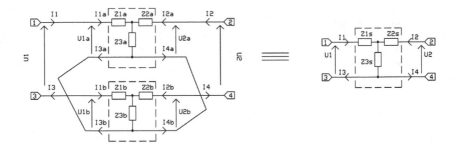

Figure 5. Basic blocks connected in series and its equivalent circuit

Two blocks configured in a series (genotype node denoted as '--') are illustrated in Fig.5. A simple analysis of this circuit leads to equations defining the impedances for the equivalent circuits:

$$Z_{1s} - Z_{1a} + Z_{1b},\tag{12}$$

$$Z_{2s} = Z_{2a} + Z_{2b},\tag{13}$$

$$Z_{3s} = Z_{3a} + Z_{3b},\tag{14}$$

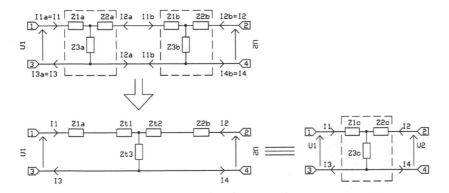

Figure 6. Basic blocks connected in cascade and its equivalent circuit

As can be seen in Fig.6, for a cascade connection (genotype node denoted as '>>') a conversion from a Π to a T-typed structure can be used for an equivalent circuit determination:

$$Z_{1c} = Z_{1a} + Z_{t1} = Z_{1a} + \frac{(Z_{2a} + Z_{1b})Z_{3a}}{Z_{\Pi ab}}, \tag{15}$$

$$Z_{2c} = Z_{2b} + Z_{t2} = Z_{2b} + \frac{(Z_{2a} + Z_{1b})Z_{3b}}{Z_{\Pi ab}}, \tag{16}$$

$$Z_{3c} = \frac{Z_{3a}Z_{3b}}{Z_{\Pi ab}}, \tag{17}$$

where:

$$Z_{\Pi ab} = Z_{2a} + Z_{3a} + Z_{1b} + Z_{3b}. \tag{18}$$

The parallel configuration (genotype node denoted as '||') is the last kind of connection assumed in (2) and it can be easily modeled by an equivalent circuit after a few transformations explained in Fig.7. Finally, impedances for an equivalent circuit can be calculated from relations:

$$Z_{1p} = \frac{Z_{1xy}Z_{3xy}}{Z_{\Pi xy}}, \tag{19}$$

$$Z_{2p} = \frac{Z_{2xy}Z_{3xy}}{Z_{\Pi xy}}, \tag{20}$$

$$Z_{3p} = \frac{Z_{1xy}Z_{2xy}}{Z_{\Pi xy}}, \tag{21}$$

where:

$$\frac{1}{Z_{1xy}} = \frac{1}{Z_{1x}} + \frac{1}{Z_{1y}} = \frac{1}{Z_{1a} + Z_{3a} + \frac{Z_{1a}Z_{3a}}{Z_{Ta}}} + \frac{1}{Z_{1b} + Z_{3b} + \frac{Z_{1b}Z_{3b}}{Z_{Tb}}}, \tag{22}$$

$$\frac{1}{Z_{2xy}} = \frac{1}{Z_{2x}} + \frac{1}{Z_{2y}} = \frac{1}{Z_{2a} + Z_{3a} + \frac{Z_{2a}Z_{3a}}{Z_{Ta}}} + \frac{1}{Z_{2b} + Z_{3b} + \frac{Z_{2b}Z_{3b}}{Z_{Tb}}}, \tag{23}$$

$$\frac{1}{Z_{3xy}} = \frac{1}{Z_{3x}} + \frac{1}{Z_{3y}} = \frac{1}{Z_{1a} + Z_{2a} + \frac{Z_{1a}Z_{2a}}{Z_{Ta}}} + \frac{1}{Z_{1b} + Z_{2b} + \frac{Z_{1b}Z_{2b}}{Z_{Tb}}},$$ (24)

$$Z_{T1xy} = Z_{1xy} + Z_{1xy} + Z_{1xy}.$$ (25)

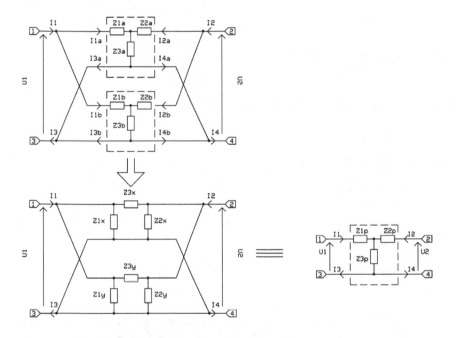

Figure 7. Basic blocks connected in parallel and the equivalent circuit

The transformations described above allow to obtain the resultant impedances Z_{1r}, Z_{2r} and Z_{3r} of the filter circuit (Fig.8) coded by a genotype tree. Finally, a frequency response of the filter loaded by the impedance Z_o can be calculated from:

$$K = \frac{U_{out}}{U_{in}} = \frac{Z_o}{(Z_o + Z_{2r})\left(1 + \frac{Z_{1r}}{Z_{3r}}\right) + Z_{1r}}.$$ (26)

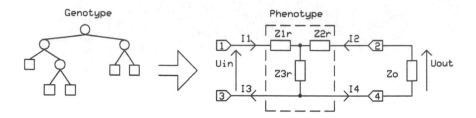

Figure 8. The resultant equivalent filter circuit

4. Genetic operations and fitness calculation

4.1. Reproduction and crossover

The first genetic operation executed after primary generation evaluation is reproduction and it completes a mating pool. A rang method of reproduction is applied in the proposed solution. This kind of operation assures the minimization of the evolutionary system's tendency to promote the average fitted phenotypes (selection pressure regulation). The probability P_{sel} of an individual selection for an intermediate pool of candidates for recombination depends on rang r value:

$$P_{sel} = P_{mn} + \left(1 - P_{mn}\right)\frac{r}{N_{mx}}, \tag{27}$$

where P_{mn} denotes an assumed minimal value of the resultant probability (to avoid the currently most fitted genotypes domination), and N_{mx} is a total number of individuals in the population. The rang r is equal to the position of an individual in population ordered by a fitness value (i.e. for the worst $r=1$ and for the best evaluated $r=N_{mx}$).

The crossover (recombination of the genetic material) process is carried out with probability P_{cr} for two parental genotypes selected randomly (with uniform probability) from a mating pool and its idea is illustrated in Fig.9. After two crossover points CP1 and CP2 random designation for genotypes (individually for each one), adequate sub-trees are exchanged between them. This process allows for preserving and propagating the genetic information of the well fitted individuals inside the generation and this operation promises to obtain better evaluated offspring genotypes. Additionally, for the proposed system, the protection against too much tree growing was applied to crossover process. Recombination is accepted only if for each offspring individual the maximal depth does not exceed the assumed D_{mx} value.

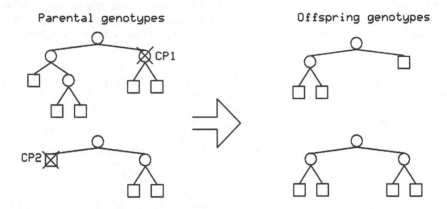

Figure 9. The idea of the crossover process

4.2. Genotypes mutations

The main goal of the mutation operations is the protection against the optimization process stagnation in the local area of the searching space. Four types of mutation procedure are used in the system described:

- a function node mutation with probability P_{1mu},
- a terminal node mutation with probability P_{2mu},
- a sub-tree deletion with probability P_{3mu},
- a sub-tree mutation with probability P_{4mu}.

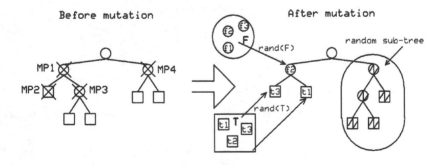

Figure 10. Genotype mutations illustration

The ideas of all these genotype modifications are illustrated in Fig.10. In case of node mutation, it is replaced by a random one (function or terminal adequately) selected from the initially assumed set (2) or (3). During the sub-tree mutation process it is deleted or replaced by
a randomly created one. Sub-tree cutting allows for simplifying the phenotype, so it is desired to the filter circuit size minimization. On the other hand, the new sub-tree adding can
lead to a genotype growing, so this process is controlled and only results with the maximal
depth up to D_{mx} are accepted.

4.3. Fitness value calculation

The quality of the phenotype is evaluated by means of fitness value calculation:

$$Q = \sqrt{\frac{\sum\limits_{m=1}^{M}\left(|K_m| - |K_{0m}|\right)^2}{w_a\big|_{a=1}^{A}}} + max\left(\frac{\left(|K_m| - |K_{0m}|\right)\big|_{m=1}^{M}}{w_a\big|_{a=1}^{A}}\right). \tag{28}$$

Equation (28) is a sum of two components. The first one is a mean square distance of the
$|K_m|$ amplitude response of the tested phenotype from the desired region of the filter de-
sign specifications $|K_{0m}|$ and it is calculated in M points totally of frequency response for
the assumed range of analysis. Besides, to assure the same impact on optimization for
each frequency band a defined in filter specifications, its distance is averaged by division
by the number w_a of frequency points included in the actually analyzed band. The second one is the average value of the maximal deviation of the frequency point for bands
and it additionally prevents the system from stagnation at average solutions. The memetic optimization system minimizes Q during evolutionary cycles and for the best phenotype this distance reaches zero.

After offspring individuals creation its terminal nodes (i.e. filter elements values) are
searched with the use of HJM in the way described in the next section and after collecting
the new generation it replaces the previous one during the succession process. Finally, the
best fitted genotype codes the structure of the filter circuit.

5. Local search

5.1. Hooke and Jeeves local search algorithm

The pattern search optimization algorithm consists of two types of moves, i.e. exploratory
move and pattern move.

The purpose of the first of them is to find and utilize information about the optimized
function values around the current base point $\mathbf{b}_{(k=0)}$ (individual). In increasing order of
indexes, each of the following circuit parameters' values is given a small increment (firstly in positive and then, if required, in the negative direction). The value of the fitness

function is checked and if there is a progress noticed, the value of this variable will be kept as a new $\mathbf{b}_{(k+1)}$ vector. This step is being repeated with reduced range of the parameters values change. When no fitness function value improvement is possible, a pattern move, starting from the current point, is made.

The aim of the pattern move is to speed up the search by information gained about the optimized function and to find the best search direction. A move from the current $\mathbf{b}_{(k+1)}$ in the direction given with $\mathbf{b}_{(k+1)}$-$\mathbf{b}_k$ is made. The fitness function value is calculated in the point given by

$$\mathbf{p}_k = \mathbf{b}_{(k+1)} - 2\mathbf{b}_k. \tag{29}$$

The procedure is continued with a new sequence of exploratory moves starting from the point $\mathbf{p}_k$. If the achieved fitness function's value is lower than the one around point $\mathbf{b}_k$, then a new base point $\mathbf{b}_{(k+2)}$ has been found. In this case a new pattern move (29) is carried out. Otherwise, the pattern move from $\mathbf{b}_{(k+1)}$ is dropped and the new exploratory sequence starts from $\mathbf{b}_{(k+1)}$. The procedure is continued until the length of the step for each of the variables has been reduced to the value assumed previously.

5.2. The implementation of the Hooke-Jeeves method

The search space in the presented research is not only discrete but also limited by the E24 sequence. Considering it several correcting procedures needed to be implemented.

5.2.1. Increasing and decreasing values of the circuit parameters

The first problem to be addressed was to properly increase the values of the circuit parameters. It has been achieved by translating the $\mathbf{V}$ and $\mathbf{O}$ sets into two vectors of integers:

$$\mathbf{V}_V = \{1,..,24\}, \tag{30}$$

$$\mathbf{V}_O = \{1,..,7\}, \tag{31}$$

where: $\mathbf{V}_v(1)$ denotes v_1=10, $\mathbf{V}_v(2)$ denotes v_2=11, etc. $\mathbf{V}_o(1)$ denotes o_1, $\mathbf{V}_o(2)$ denotes o_2, etc.

Let step_k be the current size of the variable var_i change. The i-th variable is given with:

$$var_i = \left(v_v^i, v_o^i\right), \tag{32}$$

where v_v^i is the translated value of v_e and v_o^i is the translated value of o_u (see (3)-(5)).

The increment of the var_i is carried out according to the presented algorithm:

- $v_v^i = v_v^i + step_k$

- if $v_v^i > 24$ then:

- $v_v^i = v_v^i - 24$

- $v_o^i = v_o^i + 1$

- if $v_o^i > 7$ then $v_v^i = 24 \land v_o^i = 7$

The decrement procedure is carried in a similar way, i.e.:

- $v_v^i = v_v^i - step_k$

- if $v_v^i < 1$ then:

- $v_v^i = v_v^i + 24$

- $v_o^i = v_o^i - 1$

if $v_o^i < 1$ then $v_v^i = 11 \land v_o^i = 0$

5.2.2. The „in-loop" local search

The local search procedure implemented within the evolutionary loop has been implemented in a way that the computation time was not affected significantly.

First of all, there has been a strict limit of pattern searches number set. Secondly, the initial step size has been small. It allows for exploring only a small area around the base point given with following individuals.

6. Exemplary Results

For the efficiency of the technique presentation, an automated design of the specialized filter was realized. The assumed, desired amplitude response specifications for a filter are as follows ($A=7$ bands):

$$
\begin{cases}
|K_0| \in \langle -\infty, -20 \rangle \text{dB} & \land \quad f = \langle 1e-7, 0.1 \rangle \text{MHz} \\
|K_0| \in \langle -\infty, 0 \rangle \text{dB} & \land \quad f = \langle 0.1, 0.325 \rangle \text{MHz} \\
|K_0| \in \langle -3, 0 \rangle \text{dB} & \land \quad f = \langle 0.325, 0.375 \rangle \text{MHz} \\
|K_0| \in \langle -\infty, 0 \rangle \text{dB} & \land \quad f = \langle 0.375, 0.6 \rangle \text{MHz} \quad , \\
|K_0| \in \langle -30, -20 \rangle \text{dB} & \land \quad f = \langle 0.6, 0.7 \rangle \text{MHz} \\
|K_0| \in \langle -\infty, 0 \rangle \text{dB} & \land \quad f = \langle 0.7, 0.9 \rangle \text{MHz} \\
|K_0| \in \langle -\infty, -40 \rangle \text{dB} & \land \quad f = \langle 0.9, 1 \rangle \text{MHz}
\end{cases}
\tag{33}
$$

for an assumed load resistance $Z_0=R_0=150$. The MGP system was executed with the initial parameters:

- the number of generations $G_{mx}=50$,

- the maximal allowed depth for genotypes $D_{mx}=4$,

- the population size $N_{mx}=10$,

- the crossover probability $P_{cr}=0.9$,

- the mutations probabilities $P_{mu1}=0.2$, $P_{mu2}=0.3$, $P_{mu3}=0.1$, $P_{mu4}=0.2$,

- the minimal probability of reproduction $P_{mn}=0.3$,

- the number of frequency points $M=100$,

and for the orders (5) of values for searched elements:

$$O = \{0,1, .., 6\} \ . \tag{34}$$

The best found genotype evaluated for the specifications (33) is illustrated in Fig.11 and it codes the temporary version of the filter circuit placed in Fig.12a. Next, due to radically small or high values for some elements and due to the obtained kinds of connections (e.g. shortened or opened branches, the same types of elements connected in series or in parallel), unnecessary elements from this circuit can be removed or simplified. This stage of the circuit synthesis can be supported by a simulation software tool. In Fig.12b the final filter circuit obtained for the considered example is presented. In comparison to the temporary one, its size was radically reduced without amplitude response degradation.

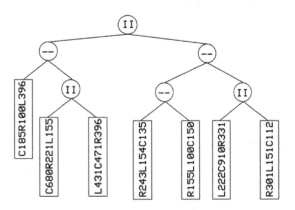

Figure 11. The resultant genotype

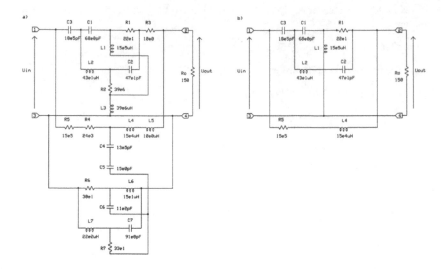

Figure 12. The structures of the designed filters: a) temporary version, b) final version

The quality of the specifications (33) keeping can be seen in Fig.13. The rectangles defining the allowed region for the filter amplitude response and simulation results obtained for the resultant circuit from Fig.12b are placed there. Only one restriction from (33) (the last frequency band) was a little violated, but the other ones are fully fulfilled. It should be emphasized, that due to discretizaton of the available values of components to a practical E24 series (4) not all theoretical shapes of frequency responses assumed on filter specifications defining stage will be practically reachable and it is necessary to conciliate with some inaccuracies. Besides, too high tolerance decreasing leads to the undesirable growing of production costs.

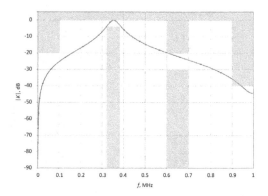

Figure 13. The amplitude response of the designed filter

7. Conclusions

The automated system for a passive filter circuits design was presented in this chapter. Any initial information about the filter circuit structure is not necessary for filter synthesis, only desired specifications should be defined. The circuit's topology as well as its elements values are optimized together in the MGP system. Thanks to the deterministic algorithm of the local searching engaging (HJM), the speed of convergence to the well evaluated solutions during the evolutionary computations grows significantly and the values of the filter's elements are adjusted to the most fitted ones for an actual circuit topology. Finally, after redundant elements elimination, the filter circuit is obtained with components selected from a practical production series (i.e. for practically reachable nominal values), and it makes the circuit realization easier. Besides, the genotypes sub-trees deletion applied in the system enables the complexity of the circuit minimization.

For the future system improving, the additional criteria for fitness calculation (28) can be added. For example, the group delay value and the kind of circuit elements (the quantity of inductors minimization) may be optimized. Besides, the applying of the models of real elements should be considered (to the parasitic parameters taken into account). Additionally, the proposed system can be adapted to the active circuits automated synthesis.

Author details

Tomasz Golonek* and Jantos Piotr

*Address all correspondence to: tgolonek@polsl.pl

Silesian University of Technology, Poland

References

[1] Koza J. R., Genetic Programming: on the programming of computers by means of natural selection, MIT Press, 1992.

[2] Banzhaf W., Nordin P., Keller R. E., Francone, F. D. Genetic Programming - An Introduction, Morgan Kaufmann Publishers Inc., San Francisco, California, 1998.

[3] Goldberg D. E., Genetic Algorithms in Search, Optimization, and Machine Learning, Kluwer Academic Publishers, Boston, 1989.

[4] Koza J. R., Bennett F. H., Andre D., Keane M. A., Dunlap F., "Automated Synthesis of Analog Electrical Circuits by Means of Genetic Programming", IEEE Transactions on Evolutionary Computation, Vol. 1, No. 2, 1997, 109-128.

[5] Tathagato Rai Dastidar, P. P. Chakrabarti, Partha Ray, "A Synthesis System for Analog Circuits Based on Evolutionary Search and Topological Reuse", IEEE Transactions on Evolutionary Computation, Vol. 9, No. 2, 2005, 211-224.

[6] Budzisz H., "Evolutional searching for circuit structures", Electronic Letters, Vol. 34, No. 16, 1998, 1543-1545.

[7] Golonek T., Jantos P., Rutkowski J., Stimulus with limited band optimization for analogue circuit testing, Metrology and Measurement Systems, Vol. XIX, No. 1, pp. 73-84, 2012.

[8] Jantos P., Rutkowski J., Evolutionary methods to analogue electronic circuits yield optimisation, Bulletin of the Polish Academy of Sciences - Technical Sciences, Vol. 56, Issue 1, 2008, pp. 9-16.

[9] Golonek T., Rutkowski J., Genetic-Algorithm-Based Method for Optimal Analog Test Points Selection, IEEE Trans. on Cir. and Syst.-II., Vol.54, No.2, 2007, pp. 117-121.

[10] Golonek T., Grzechca D., Rutkowski J., Application of Genetic Programming to Edge Decoder Design, Proc. of the Inter. Symposium on Cir. and Sys., ISCAS 2006, Greece, 4683-4686.

[11] Hooke R. , Jeeves T. A., Direct search solution of numerical and statistical problems, Assoc. Computing Machinery J., Vol.8 , No.2, 1960, 212-229.

[12] Moscato P., "On Evolution, Search, Optimization, Genetic Algorithms and Martial Arts: Towards Memetic Algorithms", Caltech Concurrent Computation Program (report 826), 1989.

[13] Land M. W. S., Evolutionary Algorithms with Local Search for Combinatorial Optimization, 1998.

[14] Ozcan E., "Memes, Self-generation and Nurse Rostering". Lecture Notes in Computer Science. Lecture Notes in Computer Science, Springer-Verlag, 3867, 2007, 85–104.

[15] Chen X., Ong Y. S., Lim M. H., Tan K. C., "A Multi-Facet Survey on Memetic Computation", Evolutionary Computation, IEEE Transactions on , vol.15, no.5, 2011, pp. 591-607.

[16] Ong Y. S., Lim M. H., Zhu N., Wong K. W., "Classification of adaptive memetic algorithms: a comparative study," Systems, Man, and Cybernetics, Part B: Cybernetics, IEEE Transactions on , vol.36, no.1, 2006, pp. 141-152.

Chapter 5

Fault Diagnosis in Analog Circuits via Symbolic Analysis Techniques

Fawzi M Al-Naima and Bessam Z Al-Jewad

Additional information is available at the end of the chapter

1. Introduction

Fault diagnosis of analog circuits has been one of the most challenging topics for researchers and test engineers since the 1970s. Given the circuit topology and nominal circuit parameter values, fault diagnosis is to obtain the exact information about the faulty circuit based on the analysis of the limited measured circuit responses. Fault diagnosis of analog circuits is essential for analog and mixed-signal systems testing and maintenance both during the design process and the manufacturing process of VLSI ASICs.

There are three dominant and distinct stages in the process of fault diagnosis: fault detection to find out if the circuit under test (CUT) is faulty comparing with the fault-free circuit, or golden circuit (This stage is usually called test in industry), fault identification to locate where the faulty parameters are inside the faulty circuit, and parameter evaluation to obtain how much the faulty parameters deviated from their nominal values and to obtain values of other circuit parameters such as branch and nodal voltages. The bottlenecks of analog fault diagnosis primarily lie in the inherited features of analog circuits: nonlinearity, parameter tolerances, limited accessible nodes, and lack of efficient models. Multiple fault diagnosis techniques are even less developed than single fault diagnosis because it is more difficult to model and detect multiple faults, particularly in the presence of tolerance or measurement noise. In addition, in multiple fault situation, one fault's effect on the circuit could be masked by the effects of the other faults. Generally speaking, there is no widely accepted paradigm for analog test or fault diagnosis even with the introduction of IEEE 1149.4 standard for mixed-signal test bus.

With recent sharp development of electronic design automation tools and widespread application of analog VLSI chips and mixed-signal systems in the area of wireless communication, networking, neural network and real-time control, the interests in analog test and fault diagnosis revives. System-on-chip solutions favored by modern electronics pose new challenges in this topic such as increased complexity and reduced die size and accessibility.

Several methods have been proposed for single fault diagnosis in linear analog circuit in the past. Multiple excitations are required and Woodbury formula in matrix theory is applied to locate the faulty parameters. This method is also applied to multiple fault diagnosis by decomposition technique assuming that each sub-circuit contains at most a single faulty parameter.

Among the different methods of fault diagnosis, the parametric fault diagnosis techniques hold an important part in the field of analog fault diagnosis. These techniques, starting from a series of measurements carried out on a previously selected test point set, given the circuit topology and the nominal values of the components, are aimed at determining the effective values of the circuit parameters by solving a set of equations generally non-linear with respect to the component values. In this chapter the role of symbolic techniques in the automation of parametric fault diagnosis of analog circuits is investigated followed by a practical numerical procedure to evaluate the faults. Being in fact the actual component values that represent the unknown quantities, fault diagnosis aims also at finding the faults locations. Symbolic approach results are particularly suitable for the automation of parametric fault diagnosis techniques [1]. Obviously all this is applicable to linear analog circuits or to nonlinear circuits suitably linearized. On the other hand, present trend is moving as much as possible to design techniques that lead to linear analog circuits, so linearity is not a so serious restriction any more [2]. It is important to note that in the analog fault diagnosis two phases can be considered: the first one is the phase of testability analysis and ambiguity group determination, while the second one is the phase of fault location and fault value determination. Testability gives theoretical and rigorous upper limits to the degree of solvability of fault diagnosis problem once the test point set has been chosen, independently of the method effectively used in fault location phase. This becomes very important in the design stage of the linear circuit in which the designer can determine the list of accessible nodes for the operator and the fault detection ability that they can provide. Concerning ambiguity groups, they can be considered as sets of circuit components that, if considered as potentially faulty, yield an undetermined system of equations. For the testability evaluation problem symbolic approach is a natural choice, because a circuit description made by means of equations in which the component values are the unknowns is properly represented by symbolic relations. Also for ambiguity group determination the symbolic approach gives excellent results [3].

For the fault location phase several different approaches can be used and all of them can be considered as an optimization problem, because, starting from measurements carried out on the CUT, the component values better fitting them have to be determined. Generally symbolic techniques are suitable for optimization problems, because the relations required by specific optimization strategies are easily generated using symbolic methods.

The aim of this chapter is to present a unified treatment of the subject with emphasis on a generalized method for multiple fault diagnosis of linear analog circuits in frequency domain. In this approach, multiple excitations and Woodbury formula are also required for fault identification. However, a recently developed ambiguity group locating technique is applied for fault identification which reduces computational cost of the test method. Multiple faults can be located directly and efficiently, thus eliminating the requirement for decomposition and the corresponding restrictions. Moreover, the methodology developed in the proposed method, (i.e., constructing fault diagnosis equation on the basis of the analysis of the fault-free circuit and the measured responses of faulty circuit, then applying the ambiguity group locating technique to identify the faulty parameters, finally evaluating all parameter values of faulty circuit exactly), can be applied to other methods developed for multiple analog fault diagnosis.

The dominant differences among these methods are the distinct fault diagnosis equations resulting from distinct circuit analysis methods and distinct excitation and measurement methods. The methods proposed in this chapter can be classified as fault verification methods under the category of Simulation-after-Test (SAT), which can provide the exact solution to the circuit parameters and can be applied to detect large parameter changes when the number of independent measurements are greater than the number of faults in the CUT.

A major improvement of these techniques is achieved through the use of symbolic techniques in formulating the fault equations and in addressing the testability problem. Furthermore a developed method for minimum size ambiguity group locating technique is used based on QR factorization and is applied to detect and identify the multiple faults. Detailed procedures for a proposed fault diagnosis program are given to help practitioners and researchers as well to grasp the basic concepts of the topic and be able to contribute to this field.

2. Basic circuit formulations

Generally, the circuit topology as well as its parameters nominal values are known in advance. Consider for example a continuous-time, time-invariant, strongly connected, linear circuit with $n+1$ nodes and p parameters. The $(n+1)^{st}$ node, denoted by zero, is assigned to be the grounded reference node while the remaining n nodes are ungrounded. All p parameters are divided into two categories: one is parameters which have admittance description such as conductance, capacitor and voltage- controlled-current source, another is parameters which have no admittance description such as impedance, inductor, current-controlled-source, operational amplifier, etc.

Of course the conventional method of analysis would be to apply the KCL to each circuit node to obtain n equations with variables being nodal voltages and parameter currents. Then constitutive equations in terms of nodal voltages and parameter currents, which define the characteristics of all parameters without admittance description, are appended to the above n KCL-based equations. The resulting system matrix from this approach would be

$$T_g X_g = W_g \qquad (1)$$

where T_g is a $g \times g$ coefficient matrix consisting of circuit parameters, X_g is a $g \times 1$ solution vector of node voltage and parameter currents, and W_g is a $g \times 1$ excitation vector composed of independent current and voltage sources, and initial conditions of capacitors and inductors. The first n rows in T_g, X_g and W_g correspond to n nodes. The resulting system equation (1) is called the *modified nodal analysis equation MNA* [4]. Note that $g=n$ for normal nodal analysis of a circuit in which all parameters have admittance description, and $g>n$ for modified nodal analysis of a circuit in which some parameters have a non-admittance description.

Traditionally, the system matrix generated from the MNA method may still have many redundant variables for analysis purposes; for example: voltages of inaccessible nodes inside sub-circuits like op-amps or currents through nonphysical branches generated from sophisticated element models. A major development step to the MNA method is to eliminate all redundant variables to generate a compacted or reduced system matrix. The reduced system matrix is formulated by programming a lookup table for every element in the network. This table has conditioned link-lists that will test which variables of the element are actually needed in the final compacted matrix and introduce the element in a way so as to eliminate the redundant variables during the formulation process. This method is termed the *compacted modified nodal analysis CMNA* [5]. Provided that the circuit functions in a stable state, the parametric values of nodal voltages and parameter currents will be finite and unique. The coefficient matrix T_g is non-singular since the circuit is a strongly connected network.

Generally the system matrix described above cannot be formulated smoothly in a computerized solution without taking the circuit topology into consideration. One important fact about circuit topology is that each parameter, say h_v ($v = 1, 2,..., p$), can be located by at most 4 circuit nodes [6]: 2 input nodes k_v and l_v and 2 output nodes i_v and j_v. For 2-terminal parameters such as resistors and capacitors, the input nodes will be the same as the output nodes: $k_v = i_v$ and $l_v = j_v$. Based on this fact, the circuit topology can be completely described by two $g \times p$ structural matrices P and Q which are defined as follows:

$$P = [p_1\ p_2\ \cdots\ p_p] = [\delta_{i_1} - \delta_{j_1}\ \delta_{i_2} - \delta_{j_2}\ \cdots\ \delta_{i_p} - \delta_{j_p}]$$
$$Q = [q_1\ q_2\ \cdots\ q_p] = [\delta_{k_1} - \delta_{l_1}\ \delta_{k_2} - \delta_{l_2}\ \cdots\ \delta_{k_p} - \delta_{l_p}]$$

$$(2)$$

where δ_v represents a $g \times 1$ vector of zeros except for the v entry, which is equal to one, and p_v and q_v represent $g \times 1$ vectors describing the locations of output nodes and input nodes, respectively. Matrices P and Q are only determined by the locations, not the values of the circuit parameters. The columns of matrix P correspond to the locations of the output nodes of circuit parameters while the columns of matrix Q correspond to the locations of the input nodes of circuit parameters.

Another important fact is that most parameters in linear circuits will enter the coefficient matrix T_g in the symbolic form

$$\begin{array}{cc} k_v & l_v \end{array}$$
$$\begin{array}{c} i_v \\ j_v \end{array} \begin{bmatrix} h_v & -h_v \\ -h_v & h_v \end{bmatrix} \tag{3}$$

with the equivalent algebraic representation being

$$\left(\delta_{i_v} - \delta_{j_v}\right) h_v \left(\delta_{k_v} - \delta_{l_v}\right)^t = p_v h_v q_v^t \tag{4}$$

where superscript t denotes transpose of matrix or vector. For any grounded node, the corresponding row or column in the symbolic form will be removed together with the δv in the algebraic form. Resistor, inductor, capacitor, dependent sources, and operational amplifier with its negative inverse gain being a parameter are examples of circuit devices described in this way. Thus the system matrix can be easily formulated using the equation

$$T_g = P diag(h) Q^t \tag{5}$$

This topological formulation allows for the automatic direct translation of the Netlist (which is the list describing the values of the circuit elements and their connections to the corresponding nodes) into circuit equations. As an example consider the circuit shown in Figure 1 following [7]. This circuit will be used later in the analysis of fault equations.

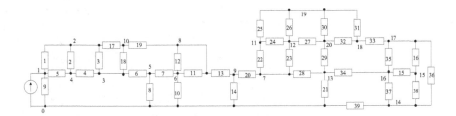

Figure 1. Example circuit

This circuit can be represented by the Netlist shown in Table 1. The first column of Table 1 is the element name and type (e.g. R for resistor, G for conductance, C for capacitor and so on). The second and third (possibly fourth and fifth depending on the element type) represent the connection nodes. Note that the ground is identified with node 0. The last column is the element value. It is possible for some elements (like a two port network or an active element model) to have more columns for its values. For the circuit in Figure 1 with unity current source J=1A applied to nodes {0,1}, it is assumed that the nominal values of the resistors are as follows (all resistors in Ω): R_1=2.125, R_2=3.6, R_3=4.7, R_4=11.5, R_5=12.6, R_6=21.2, R_7=3.7, R_8=0.54, R_9=3.54, R_{10}=3.125, R_{11}=6.6, R_{12}=5.7, R_{13}=19.5, R_{14}=12.8, R_{15}=12.2, R_{16}=3.2, R_{17}=1.54, R_{18}=8.7, R_{19}=2.27, R_{20}=3.16, R_{21}=41.7, R_{22}=31.5, R_{23}=22.6, R_{24}=51.2, R_{25}=13.7, R_{26}=3.44, R_{27}=13.4, R_{28}=31.9, R_{29}=16.1, R_{30}=11.7, R_{31}=11.5, R_{32}=17.8, R_{33}=22.2, R_{34}=23.2, R_{35}=11.4, R_{36}=18.7, R_{37}=3.12, R_{38}=33.2, R_{39}=8.67.

Element	From Node	To Node	Value
J	0	1	1
R1	1	2	2.1250
R2	2	4	3.6000
R3	2	3	4.7000
R4	3	4	11.5000
R5	1	4	12.6000
R6	3	5	21.2000
R7	5	6	3.7000
R8	0	5	0.5400
R9	0	1	3.5400
R10	0	6	3.1250
R11	6	8	6.6000
R12	6	8	5.7000
R13	8	9	19.5000
R14	0	9	12.8000
R15	15	16	12.2000
R16	15	17	3.2000
R18	3	10	8.7000
R19	8	10	2.2700
R20	7	9	3.1600
R21	0	13	41.7000
R22	7	11	31.5000
R23	7	12	22.6000
R24	11	12	51.2000
R25	11	19	13.7000
R26	12	19	3.4400
R27	12	20	13.4000
R28	7	13	31.9000
R29	13	20	16.1000
R30	19	20	11.7000
R31	18	19	11.5000
R32	18	20	17.8000
R33	17	18	22.2000
R34	13	16	23.2000
R35	16	17	11.4000
R36	14	17	18.7000
R37	14	16	3.1200
R38	14	15	33.2000
R39	0	14	8.6700

Table 1. Netlist for the circuit shown in Figure 1

It can be shown easily [4] that the node numbers in the second column of Table 1 represent the values of i in equation (2) as long as they are not zero. Similarly column 3 gives the values of j in equation (2). Thus it is very straight forward to find the P and Q matrices by tracing through the Netlist. Additionally, the system matrix can be either formulated numerically or symbolically depending on the values used for h in the diagonal matrix of equation (5). This automatic formulation procedure can be easily extended to MNA and CMNA methods with proper rubber stamps for circuit elements.

3. Conventional fault analysis methods

As discussed earlier, the conventional method for multiple fault diagnosis can be divided into three steps: fault detection, fault location determination, and finding the faulty elements values. This conventional method is readily deemed to be a numerical method by its very own nature but it is presented here as it provides basic insight to the problem and the limitations facing all numerical methods. The problem is even more complicated for multiple faults due to ambiguity presented by element tolerances not to mention that different sets of certain faults may produce very similar measured values. Further complication is present owing to the fact that only a limited number of nodes are actually accessible for measurements and testing. The conventional method will be presented without derivation which can be found in [6-8]. Despite its effectiveness in dealing with ambiguity groups, the method has several limitations:

1. The method requires multiple independent excitations among the accessible nodes. That is applying an independent source of excitation to a subset of the accessible nodes and measuring the circuit response for each source. By this, the method not only assumes that the circuit will remain linear and well-behaved under multiple excitations, it also destroyes the natural input-output relation of the circuit components and overlooks any form of signal isolation.

2. The method needs a dictionary for the behavior of the fault-free circuit under multiple excitations. The dictionary must be extensive enough to enable detecting and locating a number of simultaneous faults. Yet even when the dictionary is extensive enoguh the method may still fail in differentiating certain ambiguity groups subject to rounding errors, inaccuracies and noise that may occur in the measurements.

3. Depending on the set of accessible nodes the problem of testability and detectability of multiple faults immediately arises.

Symbolic analysis techniques aim at resolving or at least reducing some or all of these limitations thus proving to be very vital to this subject. Having that said let us begin by showing the applications of symbolic analysis techniques for multiple fault diagnosis in linear circuits.

3.1. Symbolic analysis in fault diagnosis problem

As mentioned earlier, the symbolic circuit matrix T can be easily described by the multiplication of a row operator P and a column operator Q with a diagonal matrix of symbols, where P and Q are the matrix operators (or topological matrices) that indicate location of the matrix element value in the symbolic matrix T as was given in equation (5). One can immediately notice that the diagonal matrix can be represented by a simple array while both P and Q are sparse numerical matrices. One benefit of this representation is that linear operations on rows such as addition or subtraction can be simply implemented on P while linear operations on columns can be implemented on Q without altering the diagonal symbolic matrix. For most circuit analysis applications, system equations are rarely fully dense or fully symbolic. In that respect, some matrix elements may contain not only a single symbolic element value but constant values as well. These constants do not affect the structures of matrices P and Q. In fact even if the element values are complete polynomials the representation is still not altered. Hence in solving a matrix equation like equation (1) only arrays of symbolic polynomial data structures need to be stored to represent the diagonal matrix while numerical matrices P, Q can be manipulated to solve this system using topological methods like determinant decision diagrams DDD [9] or matrix reduction methods [10].

Traditionally symbolic techniques have been used along two separate paths in fault analysis: To introduce comprehensive fault models in small and moderate sized circuits and to find the optimum set of testable components in a faulty circuit. We will describe here both techniques and show ways of utilizing them later.

3.1.1. Symbolic techniques in comprehensive fault modeling

The method described before for tracing through the Netlist to generate the P and Q topological matrices is not the only way to generate the system matrix. In fact, most of the modified and compacted methods like the tableau, MNA and CMNA methods focus on tracing through the Netlist in an element-by-element fashion, increasing the size of the generated system matrix iteratively. This approach has the advantage of providing a way to reduce the system matrix during the formulation step as will be shown shortly [5].

Consider a general admittance y connected between nodes i and j as shown in Figure 2. Assuming that the system matrix T is already generated for the other branches, the impact of this admittance (following the tableau formulation) on the system matrix is to add an additional row and column corresponding to the new system variable i_y

$$\begin{bmatrix} y_{ii} & y_{ij} & 1 \\ y_{ji} & y_{jj} & -1 \\ y & -y & -1 \end{bmatrix} \times \begin{bmatrix} V_i \\ V_j \\ i_y \end{bmatrix} = \begin{bmatrix} w_i \\ w_j \\ 0 \end{bmatrix}. \tag{6}$$

Now, if i_y is not a solution variable then it can be eliminated from the system matrix to generate a compacted matrix with respect to the axis i_y. Applying Kron's reduction to eliminate axis-3 of this matrix we get [5]

$$\begin{bmatrix} y_{ii} + y & y_{ij} - y \\ y_{ji} - y & y_{jj} + y \end{bmatrix} \times \begin{bmatrix} V_i \\ V_j \end{bmatrix} = \begin{bmatrix} w_i \\ w_j \end{bmatrix}. \tag{7}$$

Equations (6) and (7) are the conditioned stamps for the admittance y and they can be programmed into a lookup table easily.

The occurrence of faults in circuit elements generally leads to a deviation in node voltages and branch currents from nominal values. The purpose of fault diagnosis is to use voltage measurements on a limited number of nodes to verify the presence of a fault in the circuit then identify the fault location and value through simulation. However additional care must be taken since the deviations in the measured values may very well result from normal parameter tolerances.

In general a fault is generated if the nominal element value is changed from y to $y+dy$ beyond its tolerance. However instead of changing the value of y in the system matrix, owing to the symbolic approach we can simply introduce an extra faulty element dy and an extra faulty variable f as shown in Figure 3 for the passive admittance case. This deviation from nominal value will result in deviation in node voltages from V to $V+dV$ and branch currents from I to $I+I_f$ where I_f is the set of fault currents. As an example consider a linear admittance y connected between nodes i and j as shown in Figure 3.

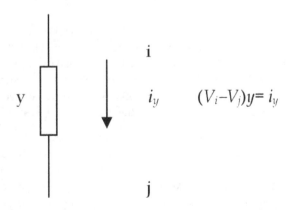

Figure 2. A general admittance example

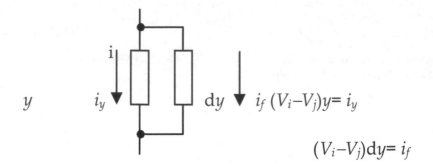

Figure 3. Fault element model for a linear admittance

Due to a fault the value of the admittance is changed to $y+dy$. Now following [5] the fault rubber stamp of (7) is changed to

$$
\begin{bmatrix}
y_{ii}+y & y_{ij}-y & 1 \\
y_{ji}-y & y_{jj}+y & -1 \\
1 & -1 & \zeta
\end{bmatrix}
\times
\begin{bmatrix}
V_i \\
V_j \\
i_f
\end{bmatrix}
=
\begin{bmatrix}
w_i \\
w_j \\
0
\end{bmatrix}
\tag{8}
$$

where we have appended the faulty element equation

$$
v_i - v_j + \zeta i_f = 0,
\tag{9}
$$

in which $\zeta = (dy)^{-1}$. We must emphasize here that this equation is appended to the original system matrix after the last nodal equation so that the faulty i_f variable will appear after the last solution variable in the fault-free system equation. The impact of this on the solution will be apparent shortly. Assuming that i_f is a solution variable, we can proceed by eliminating any non-solution variables in the stamp just like we did before to generate the faulty compacted system matrix stamp while leaving all the faulty currents as solution variables.

Clearly other fault models for the different circuit elements can also be developed by inspection. The stamping procedure for the faulty circuit elements generates the fault analysis equations. However, additional circuit elements and thus additional symbolic variables were introduced in the circuit to simulate the fault thus deeming this method suitable only for small and moderate sized networks. Assuming the original fault-free system equations were given by equation identical to (1) where T is the compacted modified system matrix, X is the chosen solution vector and W is the vector of excitation sources which might be a combination of currents and voltages. With the introduction of the faulty elements the size of the system matrix has increased. Careful consideration of element stamps like (8) show that the new faulty system equations can be written as

$$\begin{bmatrix} T & P \\ Q & R \end{bmatrix} \times \begin{bmatrix} X \\ X_f \end{bmatrix} = \begin{bmatrix} W \\ 0 \end{bmatrix} \tag{10}$$

where X is the original solution vector while X_f is the solution vector of the fault currents and voltages. This formulation resulted from the fact that the fault variable was added after the last solution variable of the fault-free circuit. Expanding this equation, it can be shown that [5]

$$(T - PR^{-1}Q)X = W \tag{11}$$

where X is now the solution vector of the faulty system. Applying Woodbury formula [5] on (11) we get

$$X = [T^{-1} + T^{-1}P(R + QT^{-1}P)^{-1}QT^{-1}]W \tag{12}$$

Expanding (12) using (10) we get

$$X = X_o + \Delta X \tag{13}$$

where

$$\Delta X = T^{-1}P(R + QT^{-1}P)^{-1}QX_o \tag{14}$$

This gives the variation in the solution vector in terms of the nominal fault-free solution vector, the topological matrices and the original system matrix. The benefit of having this variation solved symbolically is that it gives direct relationship between shifts in element values and the corresponding variation in circuit response. Once those variables are obtained symbolically it is very easy to carry out an analysis like Monte-Carlo analysis [5] to help solve the fault/tolerance ambiguity and verify the presence of a fault. Not only fault verification is possible with this equation but also locating the faulty element(s) can be done even with measurements taken from a limited set of accessible nodes using the k-fault method [11] or a linear combination matrix which will be explained later on where only a small set of the solution variables are measured to estimate the fault location.

Despite the usefulness of this approach in finding the fault model, it is highly restricted to small and moderate scale circuits. In addition, due to ambiguities, it is customary to find that the inner matrix $R+QT^{-1}P$ has become singular therefore limiting the practical use of equation (14). Nevertheless the approach is still needed to model the faults symbolically and to tackle the testability problem.

3.1.2. Symbolic solution of the testability problem

Fault diagnosis and fault location in analog circuits are of fundamental importance for design validation and prototype characterization in order to improve yield through design modification. In the analog fault diagnosis field, an essential point is constituted by the concept of testability which, independently of the method that will be effectively used in fault location, gives theoretical and rigorous upper limits to the degree of solvability of the problem, once the test point set has been chosen by the circuit designer. A well-defined quantitative measure of testability can be deduced by referring to fault diagnosis techniques of the parametric kind [12]. These techniques, starting from a series of measurements carried out on previously selected test points, are aimed at determining the upper limit of solvable circuit parameters by solving a set of equations (the fault diagnosis equations as will be shown later) which are nonlinear with respect to the component values.

The solvability degree of these nonlinear equations constitutes the most used definition of testability measure [12]. This measure can be also interpreted as an indication of the ambiguity resulting from any attempt to solve the fault equations in a neighborhood of almost any failure. In addition to being valuable for the circuit designer in determining the number of accessible nodes, it is also very important for the circuit operator since attempting to address the fault-diagnosis equation without having an estimate on the maximum number of faults that can be detected from the available test set is highly prohibited. In other words, the testability measure provides information about the number of testable components with the selected test point set. When the testability value is not at its maximum, that is when it is less than the total number of potentially faulty circuit components, the problem is not uniquely solvable and it is necessary to consider further measurements, i.e., other test points. Alternatively we can accept a reduced number of potentially faulty components in order to locate the elements which have caused the incorrect behavior of the CUT.

Generally, the second alternative is used for two reasons. First, not all the possible test points can actually be considered because of practical and economic measurement problems strictly tied with the used technology and with the application field of the circuit under consideration. Second, the number of faulty components is generally smaller than the total number of circuit components. The single fault case is the most frequent while double or triple cases are less frequent, and the case of all faulty components is almost impossible. Therefore, as the testability is normally not at its maximum, the fault diagnosis problem is dealt with by assuming the quite realistic hypothesis that the number of faulty components is bounded; that is, the k-fault hypothesis is made. Under this hypothesis, in order to locate the faulty elements with as low ambiguity as possible, it is of fundamental importance to determine a set of components that is representative of all the circuit elements. This helps reducing the solution time by providing a stopping criterion instead of wasting computer resources seeking unattainable solutions. In this section a procedure for the determination of the optimum set of testable components in the k-fault diagnosis of analog linear circuits is presented, where by the optimum set we mean a set of components representing all the circuit elements and giving a unique solution.

The procedure is based on the testability evaluation of the circuit and on the determination of the canonical ambiguity groups. Referring again, for the sake of simplicity, to parametric fault diagnosis techniques we need to make some definitions first. An ambiguity group can be defined as a set of components that, if used as unknowns (i.e., if considered as potentially faulty), gives infinite solutions during the phase of fault location determination. A canonical ambiguity group is simply an ambiguity group that does not contain other ambiguity groups. It is worth pointing out that the proposed procedure gives information independently of the method that will be effectively used in the fault location phase (both simulation after test and simulation before test methods), even if it has been developed by referring to parametric fault diagnosis techniques. Furthermore, in the automation of the procedure the use of symbolic techniques is of fundamental importance because symbolic analysis, due to the fact that it gives symbolic rather than numerical results, is particularly suitable for applications such as testability and canonical ambiguity group determination, as will be shown later.

It is necessary in this procedure to determine a set of equations describing the circuit under test and solve it with respect to the component values. In the case of analog linear time-invariant circuits, the fault diagnosis equations can be constituted by the network functions relevant to the selected test points [12] which are nonlinear with respect to the potentially faulty circuit parameters. By assuming that the faults can be expressed as parameter variations without influencing the circuit topology (as was done in the previous section where faults like short and open are not considered), the testability measure τ is given by the maximum number of linearly independent columns of the Jacobian matrix associated with the fault diagnosis equations, and it represents a measure of the solvability degree of the nonlinear fault diagnosis equations. The entries of the Jacobian matrix are rational functions depending on the complex frequency and the potentially faulty parameters. Thus, in order to evaluate the testability it is necessary to select fixed values for the potentially faulty parameter and the complex frequency. It can be shown that, once the frequency values are fixed, the rank of the obtained Jacobian matrix is constant almost everywhere, i.e., for all the potentially faulty parameter values except those lying in an algebraic variety [13]. Using this approach, the testability value, although independent of component values, is very difficult to handle and subject to round off errors if a numerical approach is used in its automation.

Generally the Jacobian matrix is very costly to find in fully symbolic form. It has been shown in [14] that starting from the network symbolic transfer functions expressed in the following way:

$$t_l(h,s) = \frac{N_l(h,s)}{D(h,s)} = \frac{\sum\limits_{i=0}^{n_l} \frac{a_i^{(l)}(h)}{b_m(h)} \cdot s^i}{s^m + \sum\limits_{j=0}^{m-1} \frac{b_j(h)}{b_m(h)} \cdot s^j}, \qquad l=1, \cdots, K \qquad (15)$$

where $h = [h_1, h_2, ..., h_p]^t$ is the vector of the potentially faulty parameters and K is the total number of equations, the testability is equal to the rank of a matrix E given by

$$
E =
\begin{bmatrix}
\partial\dfrac{a_0^{(1)}}{b_m}\big/\partial h_1 & \partial\dfrac{a_0^{(1)}}{b_m}\big/\partial h_2 & & \partial\dfrac{a_0^{(1)}}{b_m}\big/\partial h_p \\[2mm]
\vdots & \vdots & & \vdots \\[2mm]
\partial\dfrac{a_{n_1}^{(1)}}{b_m}\big/\partial h_1 & \partial\dfrac{a_{n_1}^{(1)}}{b_m}\big/\partial h_2 & & \partial\dfrac{a_{n_1}^{(1)}}{b_m}\big/\partial h_p \\[2mm]
\vdots & \vdots & & \vdots \\[2mm]
\partial\dfrac{a_0^{(K)}}{b_m}\big/\partial h_1 & \partial\dfrac{a_0^{(K)}}{b_m}\big/\partial h_2 & \cdots & \partial\dfrac{a_0^{(K)}}{b_m}\big/\partial h_p \\[2mm]
\vdots & \vdots & \cdots & \vdots \\[2mm]
\partial\dfrac{a_{n_K}^{(K)}}{b_m}\big/\partial h_1 & \partial\dfrac{a_{n_K}^{(K)}}{b_m}\big/\partial h_2 & \cdots & \partial\dfrac{a_{n_K}^{(K)}}{b_m}\big/\partial h_p \\[2mm]
\partial\dfrac{b_0}{b_m}\big/\partial h_1 & \partial\dfrac{b_0}{b_m}\big/\partial h_2 & & \partial\dfrac{b_0}{b_m}\big/\partial h_p \\[2mm]
\vdots & \vdots & & \vdots \\[2mm]
\partial\dfrac{b_{m-1}}{b_m}\big/\partial h_1 & \partial\dfrac{b_{m-1}}{b_m}\big/\partial h_2 & & \partial\dfrac{b_{m-1}}{b_m}\big/\partial h_p
\end{bmatrix}
\tag{16}
$$

This matrix is independent of the complex frequency whose entries are constituted by the derivatives of the coefficients of the fault diagnosis equations with respect to the potentially faulty circuit parameters. If the fault diagnosis equations are generated in a completely symbolic form, the testability evaluation becomes easy to perform. In this case, the entries of the matrix E can be simply led back to derivatives of sums of products and the computational errors are drastically reduced in the automation phase. Once the matrix E has been determined, testability evaluation can be performed by triangularizing E and assigning arbitrary values to the components (since as was previously mentioned, testability does not depend on component values). Yet selecting the matrix E instead of the Jacobian matrix as the testability matrix results in a different testability measure not directly related to the desired measure. However, this limitation can be overcome by splitting the fault diagnosis equation solution into two phases. In the first phase, starting from the measurements carried out on the selected test points at different frequencies, the coefficients of the fault diagnosis equations are evaluated, eventually exploiting a least-squares procedure in order to minimize the error due to measurement inaccuracy. In the second phase, the component values are obtained by solving the nonlinear system constituted by the equations expressing the previously determined coefficients as functions of the circuit parameters. In this way the following nonlinear system has to be solved:

$$\frac{a_0^{(1)}(h)}{b_m(h)} = A_0^{(1)} \quad \cdots \quad \frac{a_{n_1}^{(1)}(h)}{b_m(h)} = A_{n_1}^{(1)}$$

$$\vdots$$

$$\frac{a_0^{(K)}(h)}{b_m(h)} = A_0^{(K)} \quad \cdots \quad \frac{a_{n_K}^{(K)}(h)}{b_m(h)} = A_{n_K}^{(K)} \tag{17}$$

$$\frac{b_0(h)}{b_m(h)} = B_0 \quad \cdots \quad \frac{b_{m-1}(h)}{b_m(h)} = B_{m-1}$$

where $A_i^{(l)}$ and B_j ($i=0,\dots, n_l$, $j=0, \dots, m$-1) are the coefficients of the fault diagnosis equations in (15) which have been calculated in the previous phase. The Jacobian matrix of this system coincides with the matrix E in (16), hence, all the information provided by a Jacobian matrix with respect to its corresponding nonlinear system can be obtained from the matrix E. In particular, if rank(E) is equal to the number of unknown parameters, the component values can be uniquely determined by solving the equations in (17) through the consideration of a set of measurements carried out on the test points. If the testability τ= rank(E) is less than the number of unknown parameters R, a locally unique solution can be determined only if R-τ components are considered not faulty.

The matrix E does not give only information about the global solvability degree of the fault diagnosis problem. In fact, by noting that each column is relevant to a specific element or parameter of the circuit and by considering the linearly dependent columns of E, other information can also be obtained. For example, if a column is linearly dependent with respect to another one, this means that a variation of the corresponding component provides a variation on the fault-equation coefficients, indistinguishable with respect to that produced by the variation of the component corresponding to the other column. This means that the two components are not testable and they constitute an ambiguity group of the second order. As an example two parallel connected resistors in a circuit where we cannot distinguish which one caused the fault. By extending this reasoning to groups of linearly dependent columns of E, ambiguity groups of a higher order can be found. Then, in summary, the following definition can be formulated.

Definition 1: A set of components constitutes an ambiguity group of order j if the corresponding columns of the testability matrix E are linearly dependent. In other words, the ambiguity groups of a circuit in which a certain test point set has been chosen can be determined by locating the linearly dependent columns of the testability matrix E. Furthermore, as was mentioned, an ambiguity group that does not contain other ambiguity groups is called canonical. Therefore, a canonical ambiguity group can be defined as follows.

Definition 2: A set of k components constitutes a canonical ambiguity group of order k if the corresponding k columns of the testability matrix E are linearly dependent and every subset of this group of columns is constituted by linearly independent columns. It is important to notice that with this definition, the order of the canonical ambiguity groups cannot be greater than the testability value plus one τ +1.

In most cases the canonical ambiguity groups have some components in common. By unifying these types of groups, another ambiguity group, corresponding again to linearly dependent columns of the matrix E is obtained. We define as global an ambiguity group of the following type.

Definition 3: A set of m components constitutes a global ambiguity group of order m if it is obtained by unifying canonical ambiguity groups having at least one element in common.

Obviously, a canonical ambiguity group which does not have components in common with any other canonical ambiguity group can be considered as a global ambiguity group. Finally, the columns of the matrix E that do not belong to any ambiguity group are linearly independent. We define these as surely testable a group of components of the following kind.

Definition 4: A set of n components whose corresponding columns of the testability matrix E do not belong to any ambiguity group constitutes a surely testable group of order n.

Obviously, the number of surely testable components cannot be greater than the testability value τ, that is, the rank of the matrix E.

With these definitions in mind, the optimum set of testable components can be determined as in [12].

3.2. Formulation of fault equations

Applying equation (1) to fault-free and faulty circuits, respectively, with the same excitation sources we get

$$T_o X_o = W_o \qquad (18)$$

$$TX = (T_o + \Delta T)(X_o + \Delta X) = W_o \qquad (19)$$

where

$$T = T_o + \Delta T \qquad (20)$$

$$X = X_o + \Delta X \qquad (21)$$

It can be easily shown that

$$\Delta T X = -T_o \Delta X \qquad (22)$$

where the parameter variation can be found from the measured values of the faulty circuit, the original system matrix and an estimate of the change of the system matrix due to fault presence

$$\Delta X = -T_0^{-1} \Delta T X \qquad (23)$$

It is customary to solve equation (23) as a constrained linear optimization problem. However such an approach is limited by the solution time and ambiguity leading to local minimum convergence. Suppose that the first f of p parameters are faulty and are changed from their nominal values $h_{10}, h_{20}, ..., h_{f0}$ to new values $h_1 = h_{10} + d_1, h_2 = h_{20} + d_2, ..., h_f = h_{f0} + d_f$, where $d_1, d_2, ...,$ d_f are the parameter deviations and the deviation vector d is an $f \times 1$ vector:

$$d = [d_1 \ d_2 \ \cdots \ d_f]^t \tag{24}$$

Define F as the faulty parameter set, and assume that each faulty parameter F_v ($v = 1, 2,..., f$) is located on intersection of the corresponding rows i_v and j_v and columns k_v and l_v of the coefficient matrix T. The deviation of the coefficient matrices now has the following form:

$$\Delta T = \sum_{v=1}^{f} p_v d_v q_v^t = P_f \, diag(d) Q_f^t \tag{25}$$

where $diag$ (d) is an $f \times f$ diagonal matrix and P_f and Q_f are $g \times f$ matrices which contain 0 and ± 1 entries:

$$
\begin{aligned}
P_f &= [p_1 \ p_2 \ \cdots \ p_f] = [\delta_{i_1} - \delta_{j_1} \ \delta_{i_2} - \delta_{j_2} \ \cdots \ \delta_{i_f} - \delta_{j_f}] \\
Q_f &= [q_1 \ q_2 \ \cdots \ q_f] = [\delta_{k_1} - \delta_{l_1} \ \delta_{k_2} - \delta_{l_2} \ \cdots \ \delta_{k_f} - \delta_{l_f}]
\end{aligned}
\tag{26}
$$

Note that P_f and Q_f are sub-matrices of P and Q respectively. They can be constructed from P and Q by selecting all columns in P and Q corresponding to faulty parameters. As an example assume that there are two faulty parameters: R_9 is changed from 3.54Ω to 7.9Ω and R_{37} is changed from 3.12Ω to 2.8 Ω. The corresponding admittance deviations are $\Delta G_9 = 1/7.9 - 1/3.54 = -0.1559 / \Omega$ and $\Delta G_{37} = 1/2.8 \ 2/3.12 = 0.03663 /\Omega$. The corresponding faulty parameter set $F = [9,37]$ and the faulty nodes will be [1,14,16]. It can be easily verified that P_f and Q_f are 20×2-matrices which can be obtained from the 9th and 37th columns of the matrices P and Q respectively. It can also be verified that ΔT will have entries only at locations {1,1}, {14,14}, {14,16}, {16,14}, {16,16}.

Substituting (25) in (20) we get

$$T = T_o + P_f \, diag(d) Q_f^t \tag{27}$$

and to obtain the solution vector for the faulty circuit we use

$$X = T^{-1} W_o \tag{28}$$

It can be shown using Woodbury formula that the value of d_v ($v=1,2,..., f$) cannot be zero or infinity to meet with the requirement of inverting [6]. Since d_v being zero means fault-free parameter and only faulty parameters will be identified by following fault diagnosis algorithm, we will have only one restriction: d_v cannot be infinite, which corresponds to the case of open admittance or short impedance. But open or short faults can be dealt with by ideal switches introduced in modified nodal analysis [4]. Therefore, the proposed method can handle open and short faults as well but only if combined with a procedure that repeats the analysis of the circuit after introduction of ideal switches.

The solution vector for fault-free circuit is

$$X_0 = [x_{1,0} \ x_{2,0} \ \cdots \ x_{g,0}]^t \tag{29}$$

where subscript 0 indicates that the denoted parameters are for fault-free circuit. Hence the product of Q_f^t and X_o can be written as

$$
\begin{aligned}
Q_f^t X_0 &= [\delta_{k_1} - \delta_{l_1} \ \delta_{k_2} - \delta_{l_2} \ \cdots \ \delta_{k_f} - \delta_{l_f}]^t X_0 \\
&= [x_{k_1,0} - x_{l_1,0} \ x_{k_2,0} - x_{l_2,0} \ \cdots \ x_{k_f,0} - x_{l_f,0}]^t \\
&= [x_{k_1 l_1,0} \ x_{k_2 l_2,0} \ \cdots \ x_{k_f l_f,0}]^t
\end{aligned}
\tag{30}
$$

and it has the physical interpretation of controlling nominal signal values (e.g. voltages) on faulty parameter input terminals.

Let us define

$$
\begin{aligned}
\beta &= [\beta_1 \ \beta_2 \ \cdots \ \beta_n]^t = T_0^{-1} P_f \\
\gamma &= Q_f^t T_0^{-1} P_f = Q_f^t \beta
\end{aligned}
\tag{31}
$$

It can be shown that the deviation vector ΔX can be obtained by [6]

$$
\begin{aligned}
\Delta X &= -\beta [diag(d^{-1}) + \gamma]^{-1} Q_f^t X_0 \\
&=
\begin{bmatrix}
\alpha_{11} & \alpha_{12} & \cdots & \alpha_{1f} \\
\alpha_{21} & \alpha_{22} & \cdots & \alpha_{2f} \\
\vdots & \vdots & \cdots & \vdots \\
\alpha_{g1} & \alpha_{g2} & \cdots & \alpha_{gf}
\end{bmatrix}
\begin{bmatrix}
x_{k_1 l_1,0} \\
x_{k_2 l_2,0} \\
\vdots \\
x_{k_f l_f,0}
\end{bmatrix}
\end{aligned}
\tag{32}
$$

where

$$
\begin{aligned}
\alpha &= -\beta [diag(d^{-1}) + \gamma]^{-1} \\
&=
\begin{bmatrix}
\alpha_{11} & \alpha_{12} & \cdots & \alpha_{1f} \\
\alpha_{21} & \alpha_{22} & \cdots & \alpha_{2f} \\
\vdots & \vdots & \cdots & \vdots \\
\alpha_{g1} & \alpha_{g2} & \cdots & \alpha_{gf}
\end{bmatrix}
=
\begin{bmatrix}
\alpha_1 \\
\alpha_2 \\
\vdots \\
\alpha_g
\end{bmatrix}
\end{aligned}
\tag{33}
$$

Usually voltage measurement is easier to carry out and is less invasive to analog circuit properties than current measurement. Therefore, we only consider the use of nodal voltage measurement in this formulation. As an example γ for our example circuit of Figure 1 assuming the aforementioned faults will be given by

$$\gamma = \begin{bmatrix} 2.4475 & 0.0145 \\ 0.0145 & 2.5120 \end{bmatrix} \tag{34}$$

Once we have the fault equations formulated and the faults simulated we can proceed to fault diagnosis.

3.3. Fault diagnosis

During the fault diagnosis we have the CUT with only a limited set of accessible nodes for measurement and excitation. Suppose the ith node is accessible for measurement, then by equation (32)

$$\Delta X_i = \begin{bmatrix} \alpha_{i1} & \alpha_{i2} & \cdots & \alpha_{if} \end{bmatrix} \begin{bmatrix} x_{k_1 l_1,0} & x_{k_2 l_2,0} & \cdots & x_{k_f l_f,0} \end{bmatrix}^t \tag{35}$$

According to definition of $g{\times}f$ matrix α in equations (33) and (31), matrix α does not depend on the location of excitation sources. Thus matrix α is invariant when applying the multiple excitation method, i.e., the same coefficients α_{ij} links deviation of measurements ΔX_i and nominal signal values on faulty parameter $x_{k_j l_j}$ independent of the excitation vector applied.

After measuring the corresponding nodal voltages on the ith node with m independent excitation vectors We (e = 1, 2,..., m), we then obtain

$$\Delta X_i^{(1)} = \begin{bmatrix} \alpha_{i1} & \alpha_{i2} & \cdots & \alpha_{if} \end{bmatrix} \begin{bmatrix} x_{k_1 l_1,0}^{(1)} & x_{k_2 l_2,0}^{(1)} & \cdots & x_{k_f l_f,0}^{(1)} \end{bmatrix}^t$$
$$\Delta X_i^{(2)} = \begin{bmatrix} \alpha_{i1} & \alpha_{i2} & \cdots & \alpha_{if} \end{bmatrix} \begin{bmatrix} x_{k_1 l_1,0}^{(2)} & x_{k_2 l_2,0}^{(2)} & \cdots & x_{k_f l_f,0}^{(2)} \end{bmatrix}^t \tag{36}$$
$$\vdots$$
$$\Delta X_i^{(m)} = \begin{bmatrix} \alpha_{i1} & \alpha_{i2} & \cdots & \alpha_{if} \end{bmatrix} \begin{bmatrix} x_{k_1 l_1,0}^{(m)} & x_{k_2 l_2,0}^{(m)} & \cdots & x_{k_f l_f,0}^{(m)} \end{bmatrix}^t$$

or in matrix form

$$\Delta X_i^M = \begin{bmatrix} \Delta X_i^{(1)} \\ \Delta X_i^{(2)} \\ \vdots \\ \Delta X_i^{(m)} \end{bmatrix} = \begin{bmatrix} x_{k_1 l_1,0}^{(1)} x_{k_2 l_2,0}^{(1)} & \cdots & x_{k_f l_f,0}^{(1)} \\ x_{k_1 l_1,0}^{(2)} x_{k_2 l_2,0}^{(2)} & \cdots & x_{k_f l_f,0}^{(2)} \\ & \vdots & \\ x_{k_1 l_1,0}^{(m)} x_{k_2 l_2,0}^{(m)} & \cdots & x_{k_f l_f,0}^{(m)} \end{bmatrix} \begin{bmatrix} \alpha_{i1} \\ \alpha_{i2} \\ \vdots \\ \alpha_{if} \end{bmatrix} \tag{37}$$
$$= X_b^{MF} \alpha_i$$

where superscript M denotes the set of multiple excitations and m is the number of these excitations. The single measurement node can be one of the nodes used for multiple excitation method, and then the total number of accessible excitation nodes should be m. Assume that $f \leq m-1 \leq p$, then the coefficient matrix Xb^{MF} has more rows than columns thus guaranteeing the uniqueness of the solution to equation (37) with verification. Equation (37) establishes the linear relationship between the measured responses of the faulty circuit ΔX_i^M and the faulty parameter deviations d since vector α_i is a linear functions of d according to equation (33). Therefore equation (37) is called the *fault diagnosis equation*, and the coefficient matrix Xb^{MF} is called the *fault diagnosis matrix* [6].

As said earlier, with only a limited number of accessible nodes the issue of testability and consistency of the selected set of accessible nodes to detect f number of simultaneous faults immediately arises. However, testability is not the focus of this chapter. We assume that the given measurement set can give at least one finite solution to circuit parameters.

As the first stage of fault diagnosis, fault detection is easily implemented. If the measurement deviation vector ΔX_i^M in the fault diagnosis equation is a zero vector, obviously the CUT is judged as fault-free for the given excitation and measurement sets. Otherwise, at least one fault is judged detected by the given measurement set. To identify the faulty parameters, first let us analyze the fault diagnosis equation. The left-side of equation (37) is a known vector from measurements; the right side is the product of an unknown coefficient matrix Xb^{MF} and an unknown solution vector α_i. According to equation (30), matrix Xb^{MF} is determined by faulty parameter locations and X_0, solution vector for the fault-free circuit. Hence the columns in Xb^{MF} represent the differences between the nominal values of nodal voltages or parameter currents across the 2 input nodes of the faulty parameters. Although we do not know matrix Xb^{MF} initially for the CUT since we do not know initially the location or number of faults, but we really know all of the nodal voltages and parameter currents in the fault-free circuit!

Similarly as in equation (30), we can construct a new $m \times p$ matrix Xb^{MP} as follows

$$Q^t X_0 = [\delta_{k_1} - \delta_{l_1}\ \delta_{k_2} - \delta_{l_2}\ \cdots\ \delta_{k_p} - \delta_{l_p}]^t X_0$$
$$= [x_{k_1,0} - x_{l_1,0}\ x_{k_2,0} - x_{l_2,0}\ \cdots\ x_{k_p,0} - x_{l_p,0}]^t \tag{38}$$
$$= [x_{k_1 l_1,0}\ x_{k_2 l_2,0}\ \cdots\ x_{k_p l_p,0}]^t$$

$$X_b^{MP} = \begin{bmatrix} x_{k_1 l_1,0}^{(1)} x_{k_2 l_2,0}^{(1)} \cdots x_{k_p l_p,0}^{(1)} \\ x_{k_1 l_1,0}^{(2)} x_{k_2 l_2,0}^{(2)} \cdots x_{k_p l_p,0}^{(2)} \\ \vdots \\ x_{k_1 l_1,0}^{(m)} x_{k_2 l_2,0}^{(m)} \cdots x_{k_p l_p,0}^{(m)} \end{bmatrix} \tag{39}$$

where superscript P denotes the set of all p circuit parameters. Each column of Xb^{MP} corresponds to one circuit parameter. Evidently, the fault diagnosis matrix Xb^{MF} is a sub-matrix of Xb^{MP} and can be constructed by collecting all columns in Xb^{MP} corresponding to the faulty parameters. Apparently matrix Xb^{MF} has more rows than columns whereas Xb^{MP} has less rows than columns due to the restriction $f \leq m\text{-}1 \leq p$.

For the purpose of fault identification, we need to find out which set or sets of columns in Xb^{MP} can satisfy the fault diagnosis equation, i.e. the dependency between ΔX_i^M and the desired coefficient matrix in fault diagnosis matrix.

Basically ΔX_i^M vector for all p parameters has to be generated from the fault-free circuit and stored as a dictionary of fault-free response to m multiple excitations over the designated m accessible nodes. This dictionary will be used later to determine whether the CUT is faulty and will be used in locating the faults. It must be emphasized that only one node for voltage measurement is sufficient for this method although multiple linearly-independent excitations are required across all m accessible nodes for successful fault location. It is thus possible to use only one of the accessible m nodes to carry out the measurements while using the rest to carry out the excitations. As an example node {2} in Figure1 is selected as the only measurement node, while nodes {2, 4, 15, 16, 17} are selected as accessible nodes for the multiple excitations. That is the unit current source is applied to these nodes respectively and the corresponding nodal voltage at node {2} is measured. Thus the measured changes of nodal voltage will be

$$\Delta X^M = \begin{bmatrix} 0.89005 \\ 0.91400 \\ 0.03651 \\ 0.032306 \\ 0.038445 \end{bmatrix} \tag{40}$$

One obvious way is to have a combinatorial search through all columns in Xb^{MP}, which is the traditional way in the fault verification method [15] and requires a number of operations of the order $O\left(\sum_{i=1}^{f} \binom{p}{i} \right)$ for f limited faults among p parameters. This is equivalent to assuming that any number of faults up to f simultaneous faults have occurred randomly in any subset of the p parameters then evaluate the response to such faults and compare it to the measured response. However, the method being described here is more efficient than that and involves locating the minimum size ambiguity group which satisfies the fault diagnosis equation. An ambiguity group is defined as a set of parameters corresponding to linearly dependent columns of Xb^{MP} which in general does not give a unique solution in fault identification. Minimum size ambiguity groups (called canonical ambiguity groups) can be found using a linear combination matrix with minimum number of non-zero entries as will be shown shortly. But to generate this we need to perform a Gaussian elimination step.

3.3.1. Gaussian eliminaion step

Let us first denote an augmented $m \times (p+1)$ matrix B_S as the concatenation of the stored dictionary vector ΔX_i^M and the matrix Xb^{MP}:

$$B_S = \begin{bmatrix} \Delta X_i^M & X_b^{MP} \end{bmatrix} \qquad (41)$$

Then we will normalize the first column of matrix B_S to have a unity in its first row,

$$\hat{B}_S(i, 1) = \frac{B_S(i, 1)}{B_S(1,1)}, \quad i = 1, 2, \cdots, m. \qquad (42)$$

If the first entry of matrix B_S, $B_S(1,1)$ happens to be zero, just permute or swap the rows of B_S so that the first entry $B_S(1,1)$ is non-zero. Such a nonzero entry must exist since ΔX_i^M is a non-zero vector for faulty circuit. Eliminate the remaining entries in the first row of matrix B_S by performing a similar operation to Gaussian elimination as follows:

$$\hat{B}_S(i, j) = B_S(i, j) - \frac{B_S(i, 1)}{B_S(1,1)} B_S(1, j), \quad i = 1, 2, \cdots, m; \, j = 2,3, \cdots, p+1. \qquad (43)$$

Finally we obtain $m \times (p+1)$ matrix $\hat{B}_S$ in the following form:

$$\hat{B}_S = \begin{bmatrix} 1^{1 \times 1} & 0^{1 \times p} \\ (\Delta \hat{X}_i)^{(m-1) \times 1} & B^{(m-1) \times p} \end{bmatrix} \qquad (44)$$

where the superscript represents the size of a vector or a matrix. Matrix B is obtained from Xb^{MP} after elimination of dependence on ΔX_i^M and is called the **verification matrix** [6]. The dependency of the desired columns of matrix B surely indicates the dependency between ΔX_i^M and the desired columns of matrix Xb^{MP}. Thus we can only concentrate on the dependency among the columns of the verification matrix B.

3.3.2. QR factorization

The rank r of the matrix B determines a maximum number of faults that can be uniquely identified by solving the fault diagnosis equation. Because $m-1 < p$, B can be permuted column wise and decomposed into two linearly dependent sub-matrices as follows

$$perm(B) = \begin{bmatrix} B_1 & B_2 \end{bmatrix} = B_1 \begin{bmatrix} I & C \end{bmatrix} \qquad (45)$$

$$B_2 = B_1 C \qquad (46)$$

where *perm* refers to column-wise permutation, $(m-1) \times r$ matrix B_1 has the full column rank equal to the rank r of the matrix B, and $r \times (p-r)$ matrix C is called linear combination matrix whose columns expand a set of basis columns from B_1 into the corresponding columns of B_2. It can be easily shown that B_1 is *a sub-matrix of B with all the rows and only a subset of the col-*

umns (called the basis set) while B_2 is a sub-matrix of B with all the rows and the remaining set of the columns (called the co-basis set). Note that the selection of independent columns of B_1 is not unique and is an important issue in solving the fault diagnosis equation in the presence of ambiguities. Different partitions define different linear combination matrices C.

Since an ambiguity group is a set of circuit parameters corresponding to linearly dependent columns of B, we define a canonical ambiguity group as a minimal set of parameters corresponding to linearly dependent columns of B. This means that if any single parameter is removed from the canonical ambiguity group, then the remaining set corresponds to independent columns of B and can be uniquely solvable. A combination of canonical ambiguity groups with at least one common element was defined as ambiguity cluster.

To efficiently deal with fault verification problem, we will look for a partition (45) with the matrix C in a minimum form, which is defined as such a matrix that one or several of its columns have the maximum number of entries equal to zero. Thus, we can get the minimum number of columns in Xb^{MP} satisfying the fault diagnosis equation (37). The corresponding partition (45) is called a canonical form of the fault diagnosis equation. Notice that according to fault verification principles [15] it is enough to find a single entry in one column of C equal to zero to solve the fault diagnosis equation. Yet since many such solutions exist we will select the column with the maximum number of zeros assuming that the faulty response was caused by the smallest number of faults. This column and all rows with nonzero entries will correspond to the faulty parameters as indicated by the element of co-basis B_2 and elements of basis B_1, respectively.

One way to find these matrices from the matrix B with high numerical stability is based on QR factorization [8], which can find a solution of over determined system of linear equations that minimizes the least square error. As a result of the QR factorization of $(m-1) \times p$ verification matrix B, we obtain:

$$BE = QR \qquad (47)$$

where E is $p \times p$ column selection matrix, Q is $(m-1) \times (m-1)$ orthogonal matrix, and R is $(m-1) \times p$ upper triangular matrix. Each column of matrix E has only one nonzero entry, which is equal to one. Matrix product BE represents the permutation of the original columns of the verification matrix B requested in equation (45). Matrix R has its rank equal to the rank of matrix B. Since R is an upper triangular matrix and $m-1 < p$, R can be written as

$$R = [R_1 \quad R_2] \qquad (48)$$

where R_1 is $r \times r$ upper triangular and has its rank equal to the rank of the verification matrix B. Having this factorization computed, it can be shown that [8]

$$perm(B) = BE \qquad (49)$$

$$C = R_1^{-1} R_2 \tag{50}$$

$$B_1 = Q R_1 \tag{51}$$

Furthermore the basis set will be the row values of the non-zero elements in the first r columns of E while the co-basis will be the row values of the remaining $p-r$ columns of E. As an example, the values of R_1 for our example circuit of Figure 1 is

$$R_1 = \begin{bmatrix} 10.5475 & -2.9444 & 0.0028 & 2.3965 \\ 0 & 3.5435 & -0.0014 & -2.8011 \\ 0 & 0 & 2.4161 & -0.0005 \\ 0 & 0 & 0 & 1.7202 \end{bmatrix} \tag{52}$$

The column permutation is {$39, 15, 2, 35$, 5, 6, 7, 8, 9, 10, 11, 12, 13, 14, 3, 16, 17, 18, 19, 20, 21, 22, 23, 24, 25, 26, 27, 28, 29, 30, 31, 32, 33, 34, 4, 36, 37, 38, 1}. Thus the basis is {39, 15, 2, 35} and co-basis is {5, 6, 7, 8, 9, 10, 11, 12, 13, 14, 3, 16, 17, 18, 19, 20, 21, 22, 23, 24, 25, 26, 27, 28, 29, 30, 31, 32, 33, 34, 4, 36, 37, 38, 1}.

3.3.3. Swapping performance

A single QR run cannot guarantee that the matrix C will be obtained with one or several of its columns having the maximum number of zero entries if the proper basis is not selected. To find the minimum form partition, we have to swap one parameter of the basis with one parameter of the co-basis in the ambiguity cluster in order to increase number of nonzero entries in C. Note that swapping parameters of the basis and the co-basis can be performed independently in different ambiguity clusters, since different clusters have mutually disjoint sets of parameters. There are simply two conditions to consider in swapping performance:

a. The necessary condition for swapping to increase the number of zero entries in C is that the columns of basis and co-basis to be swapped have a singular 2×2 sub-matrix of non-zero entries.

Let us consider a linear combination matrix C with a 2×2 singular sub-matrix

$$R_1 = \begin{bmatrix} 10.5475 & -2.9444 & 0.0028 & 2.3965 \\ 0 & 3.5435 & -0.0014 & -2.8011 \\ 0 & 0 & 2.4161 & -0.0005 \\ 0 & 0 & 0 & 1.7202 \end{bmatrix} \tag{53}$$

with all nonzero entries. If we swap the jth element of the basis with k^{th} element of the co-basis, then after swapping, the kth column of C changes to

$$C_k = -\frac{1}{c_{jk}}[c_{1k} \quad c_{2k} \quad \cdots \quad 1 \quad \cdots \quad c_{rk}]^t \tag{54}$$

In addition, all other columns of matrix C will be equal to

$$C_n = \left[c_{1n} - \frac{c_{jn}c_{1k}}{c_{ik}} \quad c_{2n} - \frac{c_{jn}c_{2k}}{c_{ik}} \quad \cdots \quad \frac{c_{jn}}{c_{ik}} \quad \cdots \quad c_{rn} - \frac{c_{jn}c_{rk}}{c_{ik}}\right]^t \tag{55}$$

Such that all zero locations in the kth column of C will be zero as they were in the original C. However, as can be deducted from (53), a nonzero location c_{im} in row i and column m will become zero. It is understood that if one element in the current basis has been swapped into the basis by the previous swapping performance, then this element will not be considered during the later swapping.

Any columns of C with zero entries form an ambiguity group F and has to be considered for further processing. Since ambiguities may exist in the original matrix Xb^{MP} then F contains all faults in the CUT only if the corresponding columns in Xb^{MP} are independent. Hence we must consider the following condition

b. The necessary condition for an ambiguity group F of the linear combination matrix C to contain the set of all faults in the tested circuit is that the rank of the corresponding columns in matrix X_b^{MP} is equal to the cardinality of F

$$rank(\text{columns in } X_b^{MP} \text{corresponding to } F) = card(F) \tag{56}$$

Thus according to this condition any ambiguity group of the verification matrix which do satisfy (55) needs to be further analyzed. The stopping criterion for the above procedure can simply be τ the testability measure found from the symbolic analysis.

As an example for our circuit of Figure 1, careful study of the generated C matrix reveals multiple zero entry at columns 5,12, 32, 33, and 34 (corresponding to nodes {9}, {16}, {36}, {37}, and {38} from the co-basis). The non-zero row entries will either be on rows 1, 2, 4 (corresponding to nodes {39}, {15} and {35} from the basis) or on rows 2, 3, and 4 (corresponding to nodes {15}, {2} and {35}). Thus the corresponding ambiguity clusters include {39, 15, 9, 35}, {16, 15, 2, 35}, {39, 15, 36, 35}, {39, 15, 37, 35}, and {39, 15, 38, 35}. Yet none of these ambiguity clusters satisfies condition (b) except for the first one. Accordingly only one suspicious faulty group F={39, 15, 35, 9} is qualified with parameter {9} from the co-basis and parameters {39, 15, 35} from the basis. The current minimum size of qualified F is 4.

Searching for the 2×2 singular matrix with non-zero entries in C reveals that parameter {9} from the co-basis should be swapped with the parameter {39} from the basis according to the swapping procedure in condition (a), and a new matrix C results. Re-applying condition (b) to the new matrix C, 5 qualified suspicious faulty groups are obtained: F={9, 2, 35, 5}, F={9, 15, 35, 39}, F={9, 15, 35, 36}, F={9, 15, 38} and F={9, 37}. Obviously, F={9, 37} is the unique solution with the minimum size equal to 2. Since no smaller size of faulty set F can be found by swapping, thus F={9, 37} is the only solution located by the procedure of fault diagnosis which is the exact solution for the given CUT.

Once the fault locations are determined the fault values must be evaluated and compared with the element tolerances before giving a final judgment on the circuit whether being faulty or not.

3.3.4. Parameter evaluation

After locating the faulty parameters, the matrix Xb^{MF} can be found from the matrix Xb^{MP} by taking only the columns corresponding to the fault locations. Then the invariant vector α_i can be uniquely solved from equation (37)

$$\alpha_i = \left(\left(X_b^{MF}\right)^t X_b^{MF}\right)^{-1}\left(X_b^{MF}\right)^t \Delta X_i^M \qquad (57)$$

where this form is used since the system is over-determined with Xb^{MF} being non-square. Finally the deviation vector d can be exactly computed by

$$d = \alpha_i ./ \left(\beta - \alpha_i \gamma\right) \qquad (58)$$

where ./ is an element-by-element division of two vectors. What remains after evaluating the deviations is to compare them to the element allowable tolerances to decide finally whether the measured CUT performance is still considered acceptable or deemed faulty. Basically, when the fault locations and parameter deviations are found, all the circuit can be re-solved to get all the node voltages and element currents of the CUT.

3.4. Mixed symbolic numerical algorithm for fault diagnosis

A computer program which implements the fault diagnosis discussed above can be easily advised. In Phase 1, a topological description of the circuit is obtained and the circuit is solved numerically. Since nominal values of circuit parameters are known, all nodal voltages in fault-free circuit can be solved by (18). In phase 2, an upper limit for the testability τ needs to be determined for the provided set of accessible nodes. It is not generally required to obtain a fully symbolic solution of the circuit and only a partial symbolic solution would be sufficient. In phase 3 we need to measure the nodal voltages of the ith node in the CUT under multiple excitation method to obtain measurement deviation vector ΔX_i^M. In phase 4 we need to generate the fault locator matrix Xb^{MP} from equations (38) and (39) then use it to find the linear combination matrix C after the Gaussian elimination step and the QR factorization. In Phase 5, analysis of the combination matrix is done where F denotes one suspicious fault set and min(size(F)) represents a scalar which is equal to the minimum size of all suspicious fault sets. In Phase 6, if several suspicious fault sets have the same minimum size, min(size(F)), select one of them arbitrarily for analysis. Only one parameter in the selected F is from the co-basis and the remaining parameters are from the basis. Swap that co-basis parameter which corresponds to column k in matrix C with one of basis parameters which corresponds to row j in the matrix C. By (53) and (54), all zero entries in the column k of matrix C will be held after swapping while new zero-entry will appear in another column of new matrix C, thus the new value of min(size(F)) will be equal to, or less than the old value before swapping.

There are two rules for swapping. One is that row j is selected with nonzero c_{jk} on the intersection of row j and column k of matrix C. Another rule is that if one parameter in the current basis has been swapped into the basis by the previous swapping operation, then this element will not be considered during the later swapping operation. Usually m-1 is far less than p, and the rank of $r×(p-r)$ matrix C, r is not greater than m-1, thus there are far less basis parameters than co-basis parameters. The comprehensive swapping between the co-basis parameter k and the basis parameters are very limited, as a result of the two swapping principles.

In Phase 7 equivalent adjoint suspicious fault sets are recorded. In Phase 8 the corresponding fault diagnosis matrices Xb^{MF} are found from the fault locator matrix. In Phase 9 the invariant vector α_i is evaluated. Phase 10 is used for verification. One or several suspicious fault sets with minimum size are used to compute the deviation vector ΔX. If a computed vector matches the real measured vector ΔX_i^M, the corresponding fault set F is our final solution to faulty parameters. Otherwise, we discard this set, and turn to the adjoint suspicious fault sets recorded in Phase 7. Verification in this phase continues until at least one qualified solution to faulty parameters is found. Otherwise, the CUT is concluded as un-solvable because the restriction $f \le m - 1$ is not satisfied. In the final Phase the parameter variations are compared to the element tolerances to decide if the circuit response is indeed faulty or just shifted within the accepted tolerance.

4. Conclusion

In this chapter, a generalized fault diagnosis and verification approach for linear analog circuits was discussed. Fault verification methods intend to obtain the information about the faulty circuit based on the limited measured responses of the faulty circuit. There are two easily implemented prerequisites: one is that the circuit topology and nominal values of circuit parameters should be known, another is that the number of measurements minus one is not less than the number of faulty parameters. A symbolic method is proposed to solve the testability problem during the detection, and location of the multiple faults in a linear analog circuit in frequency domain, then to exactly evaluate the faulty parameter deviations.

Applying the Woodbury formula in the matrix theory to the modified nodal analysis, fault diagnosis equation is constructed to establish the relationship between the measured responses and the faulty parameter deviations in a linear way. A numerically robust approach has been modified to fit the condition stated in this chapter in order to implement fault location, i.e., location of the minimum size ambiguity group in the fault diagnosis equation based on QR factorization. Parameter evaluation is then performed from results of the analysis of fault diagnosis equation.

One node for voltage measurement is sufficient for the proposed method although multiple excitations are required for fault location. Although the faulty parameter deviation cannot be infinity, open or short condition can be dealt with well by switches in modified nodal analysis.

Therefore, the faults can be parametric or catastrophic. The proposed method is extremely effective for large parameter deviations and a very limited number of accessible nodes used for excitations and measurements. The computation cost for the fault location is on the order of $O(p^3)$, and compares favorably with the combinatorial search traditionally used in fault verification methods which requires the number of operations $O\left(\sum_{i=1}^{f}\binom{p}{i}\right)$.

Author details

Fawzi M Al-Naima[1*] and Bessam Z Al-Jewad[2]

*Address all correspondence to: fawzi.alnaima@ieee.org

1 College of Engineering, Nahrain University, Baghdad, Iraq

2 Dept. of Communications and Computers Engg., Cihan University, Erbil, Iraq

References

[1] Fedi G, Giomi R, Luchetta A, Manetti S, Piccirilli M C. On the Application of Symbolic Techniques to the Multiple Fault Location in Low Testability Analog Circuits. IEEE Transaction on Circuits and Systems–II 1998; 45(10), 1383-1388.

[2] Bushnell M L, Agrawal V D. Essentials of Electronic Testing. Kluwer Academic Publisher; 2000.

[3] Fedi G, Giomi R, Luchetta A, Manetti S, Piccirilli M C. Symbolic Algorithm for Ambiguity Group Determination in Analog Fault Diagnosis: proceedings of Europian Conference on Circuit Theory and Design, ECCTD1997, Budapest, Hungary, August 1997.

[4] Vlach J, Singhal K. Computer Methods for Circuit Analysis and Design. Van Nostrand Reinhold; 1994.

[5] Al-Naima F M, Al-Jewad B Z. Mixed Symbolic-Numerical Techniques in Fault Diagnosis Using Fault Rubber Stamps. IEEE International Conference on Electronics, Circuits and Systems, ICECS2011, Beirut, December 2011.

[6] Liu D, Starzyk J A. A Generalized Fault Diagnosis Method in Dynamic Analogue Circuits. International Journal of Circuit Theory and Applications ICTACV, 2002; 30(5) 487-510.

[7] Starzyk J A, Liu D. Multiple Fault Diagnosis of Analog Circuits by Locating Ambiguity Groups of Test Equation. IEEE International Symposium on Circuits and Systems, ISCAS 2001.

[8] Starzyk J A, Manetti S, Fedi G, Piccirilli M C. Finding Ambiguity Groups in Low Testability Analog Circuits. IEEE Transactions on Circuits and Systems–I: Fundamental Theory and Applications 2000; 47(8).

[9] Sanchez-Lopez C, Fernandez F V, Tlelo-Cuautle E, Tan S X -D. Pathological Element-Based Active Device Models and Their Application to Symbolic Analysis. IEEE Transactions on Circuits and Systems I: Regular papers, vol. 58, pp. 1382–1395, June 2011.

[10] Brown H E. Solution of Large Networks by Matrix Methods. John Wiley and Sons; 1985.

[11] Liao W, Liu J. Research on k-Fault Diagnosis and Testability in Analog Circuit. WSEAS Transactions on Electronics, 2008; 5(10).

[12] Fedi G, Piccirilli M C, Starzyk J. Determination of an Optimum Set of Testable Components in the Fault Diagnosis of Analog Linear Circuits. IEEE Transactions on Circuits and Systems–I: Fundamental Theory and Applications, 1999; 46(7).

[13] Sen N, Saeks R. Fault Diagnosis for Linear Systems via Multifrequency Measurement. IEEE Transactions on Circuits and Systems, 1979; CAS-26 457–465.

[14] Liberatore A, Manetti S, Piccirilli M C. A New Efficient Method for Analog Circuit Testability Measurement: proceedings of IEEE Instruments Measurement Technology Conference, IMTC1994, Hamamatsu, Japan, May 1994.

[15] Bandler J W, Salama A E. Fault Diagnosis of Analog Circuits: proceedings of the IEEE, 1985; 73(8) 1279-1325.

Permissions

The contributors of this book come from diverse backgrounds, making this book a truly international effort. This book will bring forth new frontiers with its revolutionizing research information and detailed analysis of the nascent developments around the world.

We would like to thank Yuping Wu, for lending his expertise to make the book truly unique. He has played a crucial role in the development of this book. Without his invaluable contribution this book wouldn't have been possible. He has made vital efforts to compile up to date information on the varied aspects of this subject to make this book a valuable addition to the collection of many professionals and students.

This book was conceptualized with the vision of imparting up-to-date information and advanced data in this field. To ensure the same, a matchless editorial board was set up. Every individual on the board went through rigorous rounds of assessment to prove their worth. After which they invested a large part of their time researching and compiling the most relevant data for our readers. Conferences and sessions were held from time to time between the editorial board and the contributing authors to present the data in the most comprehensible form. The editorial team has worked tirelessly to provide valuable and valid information to help people across the globe.

Every chapter published in this book has been scrutinized by our experts. Their significance has been extensively debated. The topics covered herein carry significant findings which will fuel the growth of the discipline. They may even be implemented as practical applications or may be referred to as a beginning point for another development. Chapters in this book were first published by InTech; hereby published with permission under the Creative Commons Attribution License or equivalent.

The editorial board has been involved in producing this book since its inception. They have spent rigorous hours researching and exploring the diverse topics which have resulted in the successful publishing of this book. They have passed on their knowledge of decades through this book. To expedite this challenging task, the publisher supported the team at every step. A small team of assistant editors was also appointed to further simplify the editing procedure and attain best results for the readers.

Our editorial team has been hand-picked from every corner of the world. Their multi-ethnicity adds dynamic inputs to the discussions which result in innovative outcomes. These outcomes are then further discussed with the researchers and contributors who give their valuable feedback and opinion regarding the same. The feedback is then collaborated with the researches and they are edited in a comprehensive manner to aid the understanding of the subject.

Apart from the editorial board, the designing team has also invested a significant amount of their time in understanding the subject and creating the most relevant covers. They scrutinized every image to scout for the most suitable representation of the subject and create an appropriate cover for the book.

The publishing team has been involved in this book since its early stages. They were actively engaged in every process, be it collecting the data, connecting with the contributors or procuring relevant information. The team has been an ardent support to the editorial, designing and production team. Their endless efforts to recruit the best for this project, has resulted in the accomplishment of this book. They are a veteran in the field of academics and their pool of knowledge is as vast as their experience in printing. Their expertise and guidance has proved useful at every step. Their uncompromising quality standards have made this book an exceptional effort. Their encouragement from time to time has been an inspiration for everyone.

The publisher and the editorial board hope that this book will prove to be a valuable piece of knowledge for researchers, students, practitioners and scholars across the globe.

List of Contributors

Soumyasanta Laha and Savas Kaya
School of Electrical Engineering & Computer Science, Ohio University, Athens, OH, USA

Tales Cleber Pimenta, Gustavo Della Colletta, Odilon Dutra, Paulo C. Crepaldi, Leonardo B. Zocal and Luis Henrique de C. Ferreira
Universidade Federal de Itajuba-UNIFEI, Brazil

Zygmunt Garczarczyk
Silesian University of Technology, Gliwice, Poland

Tomasz Golonek and Jantos Piotr
Silesian University of Technology, Poland

Fawzi M Al-Naima
College of Engineering, Nahrain University, Baghdad, Iraq

Bessam Z Al-Jewad
Dept. of Communications and Computers Engg., Cihan University, Erbil, Iraq

Printed in the USA
CPSIA information can be obtained
at www.ICGtesting.com
JSHW011327221024
72173JS00003B/78